T0135488

Florian Doleschal

Perception of Vehicle Interior Sounds with Electrified Drives

Measurements and Pleasantness Estimations using a Long Short-Term Memory Model

Logos Verlag Berlin

 λογος

Bibliographic information published by the Deutsche Nationalbibliothek

The Deutsche Nationalbibliothek lists this publication in the Deutsche
Nationalbibliografie; detailed bibliographic data are available
on the Internet at http://dnb.d-nb.de

ISBN 978-3-8325-5590-0

Logos Verlag Berlin GmbH
Georg-Knorr-Str. 4, Geb. 10,
D-12681 Berlin
Germany

Tel.:   +49 (0)30 / 42 85 10 90
Fax: +49 (0)30 / 42 85 10 92
http://www.logos-verlag.de

UNIVERSITÄT
MAGDEBURG

FAKULTÄT FÜR
MASCHINENBAU

Perception of Vehicle Interior Sounds with Electrified Drives: Measurements and Pleasantness Estimations using a Long Short-Term Memory Model

Dissertation
zur Erlangung des akademischen Grades

**Doktoringenieur
(Dr.-Ing.)**

von M. Sc. Florian Doleschal
geb. am 20.09.1992 in Offenbach am Main
genehmigt durch die Fakultät für Maschinenbau
der Otto-von-Guericke-Universität Magdeburg

Gutachter:
Prof. Dr.-Ing. Hermann Rottengruber
Prof. Dr. rer. nat. Jesko L. Verhey

Promotionskolloquium am 21.12.2022

I

Acknowledgements

I would like to thank the following people supporting me in writing this dissertation:

Firstly, I would like to thank my supervisors Prof. Dr. Hermann Rottengruber and Prof. Dr. Jesko Verhey. Their immense knowledge about various fields of acoustics enabled them to provide invaluable feedback for bringing this work to a higher level and for sharpening the focus of the work to the relevant points.

Secondly, I would like to thank my colleagues Martin Gottschalk and Monika Kordus and the research assistants Nour Alrefai, Benjamin Beyer, Joshua Birenheide, Patrick Blobel, Frederike Giese, Christoph Leugers and Anne Kunert for proofreading my dissertation and conducting more than 1000 hours of auditory experiments, which are a fundamental basis of this work.

Furthermore, I would like to thank my family, my parents Sabine and Peter and my sisters Laura and Selina Doleschal. At last, I would like to thank my girlfriend Xueling Sun, who provided a lot of emotional support to me during all the years of working on that dissertation.

Kurzfassung

Das Innengeräusch ist ein wesentliches Kriterium beim Kauf eines neuen oder gebrauchten Fahrzeugs. Sowohl bei reinelektrischen Fahrzeugen, als auch bei solchen mit Hybridantrieben können tonale Geräuschanteile die Angenehmheit des Innengeräuschs maßgeblich beeinflussen. Bisherige Studien zeigten, dass neben den Reifen-/Fahrbahn- und Windgeräuschen insbesondere hochfrequente Geräuschanteile die Angenehmheit reduzieren. Die Hörbarkeit dieser Komponenten wurde mittels einer angepassten Version eines auditorischen Verdeckungsmodells bestimmt. Tonale Komponenten sind insbesondere während transienter Fahrzustände hörbar, weshalb eine dynamische Betrachtung der psychoakustischen Einflussgrößen auf die Angenehmheit erforderlich ist. Der Modellansatz eines Long Short-Term Memory (LSTM)-Netzwerks ermöglicht hierbei die Modellierung der Angenehmheit als Einzahlwert in Abhängigkeit von zeitabhängigen psychoakustischen Parametern. Die Angenehmheit wurde in zahlreichen Hörversuchen erhoben, wobei sowohl Originalaufnahmen aus dem Fahrzeuginnenraum, als auch Geräusche mit gezielten Pegelveränderungen einzelner Geräuschkomponenten und solche mit hinzugefügten synthetischen Komponenten verwendet wurden. Das übergeordnete Ziel der Dissertation ist, ein Vorhersagemodell für die Angenehmheit von Fahrzeuginnengeräuschen von Fahrzeugen mit elektrifizierten Antrieben zu entwickeln. Die Vorhersagewerte können als Grundlage für die Entwicklung von entsprechenden aktiven und passiven Geräuschverbesserungsmaßnahmen dienen.

Abstract

The interior sound of vehicles is a major criterion when buying a new or used vehicle. For both pure-electric and hybrid vehicles, separately audible tonal components severely influence the pleasantness of the interior sound. Previous studies revealed that for sounds of vehicles with electrified drives, besides the tire-road and wind noise components, mainly higher-frequency tonal components influence the pleasantness. Within the present work, the audibility of those components has been determined by a modified version of an auditory masking model. For those vehicles, disturbing sound components are usually prominent during transient driving conditions. Therefore, the temporal changes of psychoacoustic parameters should be considered for the pleasantness prediction. A long short-term memory neural network depicts the relationship between time series of psychoacoustic parameters and a single value of pleasantness, which has been acquired by conducting auditory experiments with both original and augmented electric and hybrid vehicle interior sounds. The general scope of the dissertation is to develop a model of pleasantness for interior sounds of vehicles with electrified drives using a long short-term memory model approach. The predictions form the basis for constructive countermeasures or active sound design concepts to improve the pleasantness of the interior sound.

Table of Contents

List of Abbreviations

AFC...alternative forced choice
biLSTM .... bidirectional long short-term memory
BTL ....Bradley, Terry, Luce
cd ....coast-down
CNN .... convolutional neural network
ERB$_N$....................................normalized equivalent rectangular bandwidth
fl...full-load run-up
HL ....Hearing Level
HMS..Hearing Model of Sottek
LSTM .... long short-term memory
MOTC .... magnitude of tonal content
OTPA ....operational transfer path analysis
PWM ....pulse width modulation
RMSE .... root-mean-square error
ru ....run-up
SPL.. sound pressure level
SVM .... support-vector machine

1 Introduction

The pleasantness of the interior sound is a major criterion for the purchase decision of a new or used vehicle. Specifically, the audibility of tonal components in the interior of vehicles with electrified drives (pure-electric and hybrid) is the major challenge in the field of interior sound research and active sound design measures. For these vehicles with electrified drives, the sounds are typically observed in transient driving conditions, such as run-up and coast-down conditions, and considerably reduce the perceived pleasantness. At constant driving conditions, the emission of tonal components is low. A common approach is to assess the perception of these interior sounds with jury evaluations, which are time-consuming and therefore costly. Thus, the development of a pleasantness assessment model for vehicles with electrified drives is desirable. It is reasonable to assume that pleasantness is based on psychoacoustic parameters. In contrast to previously developed models for vehicles with combustion engines, the dynamic changes of the psychoacoustic parameters are particularly relevant and therefore have to be considered during the development process of the pleasantness assessment model.

Thus, the main scope of this dissertation is to develop a dynamic pleasantness model based on psychoacoustic sensations and to provide suggestions for an active sound concept for vehicles with electrified drives.

The second chapter of the thesis provides a brief overview of the current state of research. This includes an introduction to the different types of noise, which are commonly audible in the interior of vehicles with electrified drives. Since some of the experiments are based on separated sound components, a separation algorithm is introduced that allows for a separation of recorded vehicle interior sounds into their components and allocates the components to the emitting sound sources. Finally, a short overview of the previously developed models is provided, which includes both equation-based regression and machine-learning approaches.

The third chapter describes three different approaches to estimate the audibility of sound components. This is a key element of the analysis of tonal components since only audible components contribute to the pleasantness. The first one is the experimental approach using an alternative forced choice (AFC) method. This approach is very time-consuming.

1

Hence, in the other approaches, the audibility is estimated using model simulations. In the first model approach, the AFC experiment is simulated by using the model as an artificial observer. Since those simulations are relatively time-consuming and require a high computational power, which might hamper the practical usage of the simulation, a second model approach is presented for tonal components that is based on the critical masking ratio of the tonal component to the relevant background noise.

The fourth chapter describes the preliminary experiments, which were conducted to gain first insights into the relevant parameters and possible measures to actively improve the pleasantness of vehicle interior sounds. To this end, artificial stimuli were generated, which contain key elements of vehicle interior sounds, i.e. tonal components and background noise. The use of artificial sounds allows for full control of the stimulus parameters. Several studies indicated that the magnitude of tonal content (MOTC, German translation "Tonhaltigkeit") has a significant influence on pleasantness. Since the currently available models to assess the MOTC sometimes provide contradictory results [1], the perceived MOTC was additionally evaluated for all investigated stimuli. The results of the preliminary experiments formed the basis for the selection of the stimuli and the design of the subsequent experiments with recorded sounds and revealed the suitability of different active sound design measures for recorded stimuli.

The fifth chapter shows the experiments of the recorded stimuli. Together with the experiments with augmented sounds described in the sixth chapter, the recorded stimuli provide the time-variant psychoacoustic parameters as inputs and the single-valued pleasantness values as targets used for the development process of the pleasantness assessment model.

The sixth chapter describes the data augmentation approach to increase the number of stimuli. The machine-learning approach to estimate pleasantness requires a high number of stimuli, which is beyond the scope of this measurement campaign. Therefore, the separated sound components were used to either increase or decrease the level of different sound components and to conduct controlled variations of certain stimulus parameters. The separated sound components were further used to generate subharmonics, whose frequencies were set relative to the frequencies of the related sound components using a vocoder-based approach. Besides the increase of the dataset, the experimental results reveal the influence of different real and artificial sound components on the variables pleasantness and MOTC and demonstrate the applicability of active sound design measures.

The seventh chapter describes the development process of the long short-term memory model (LSTM), which includes the calculation of the time-varying psychoacoustic parameters and the data preprocessing. Afterward, the architecture of the model and its estimation is described in detail. The validation process as a part of the estimation process ensures the applicability of the model to sounds, which were not used for the training process of the model. Furthermore, a temporal analysis was conducted to evaluate how relevant certain time segments of the input signals are for the pleasantness perception.

The last chapter of the work draws a conclusion about the applicability of the developed model and provides an outlook on the relevant measures, which should be carried out to improve the pleasantness of the interior sound based on the experimental results and the pleasantness estimation of the model.

2 State of Research

The first section of this chapter describes the sound components in the vehicle interior, which include both the sound components from the electric and combustion powertrain and the tire-road and wind noise. The second section shows the sound separation and allocation algorithm developed by Fröhlingsdorf and Pischinger, 2022 [2], which was used to split up the vehicle interior sound into individually listenable components and to allocate them to the emitting sources. The third section provides a brief overview of the fundamentals of the pleasantness and magnitude of tonal content (MOTC) and relevant works regarding the perception of vehicle sounds. The last section of this chapter gives an overview of the field of active sound design, where the outcomes of this work could be used to increase the pleasantness of the vehicle interior sound.

2.1 Interior Sound Components of Vehicles with Electrified Drives

Electric Motor

One of the main tonal sound sources in the vehicle of pure-electric and hybrid vehicles is the electric motor. In contrast to combustion engines, the electric motor does not emit any sound in during idle and barely emits any sound during cruise conditions. However, the emissions from the electric motor during part-load and full-load acceleration and during recuperation in coast-down conditions could severely affect the driving comfort in the vehicle interior. Due to the relatively high frequencies, engine encapsulations are commonly used to quieten the emissions from the electric motor [3].

The noise emissions from the electric motor are commonly referred to as motor orders. Rotating systems can emit tonal components, which are multiples of the rate of rotation of the system. As an example, electric motor orders commonly emit the 48th order, meaning that the frequency of the emitted tone is 48 times higher than the rate of rotation of the motor. Electric motors in passenger vehicles emit different types of orders, which can be grouped according to the underlying physical principles [4]:

- Radial magnetic forces
- Tangential magnetic forces
- Structural resonances

5

Generally, the electric motor of a vehicle is either a synchronous or an asynchronous motor. In both cases, the combination of pole pairs and slots has an influence on the harmonics of exciting orders, even though the excited orders differ between both types. The radial magnetic forces could be further distinguished into several types. The largest acoustic impact results from the so-called breathing mode causing radial vibrations of the stator core. The other two effects are of minor relevance compared to the breathing mode: The bending mode describes the effect that radial magnetic forces induce a one-sided magnetic pull on the rotor, while the higher-order modes cause wave-shaped deflections of the rotor [4, 5].

The excitation due to tangential magnetic forces is generally lower than due to radial forces. Tangential magnetic forces are mainly relevant for permanent magnet machines and occur due to periodic changes in the magnetic flow density. The changes in the magnetic flow density result from the time-varying relative positions of the rotor and the stator slots. The resulting audible motor orders are mainly relevant for low rotational speeds of the electric motor [5].

The engine structure can either amplify or dampen vibrations due to magnetic forces [5]. The amplifications are frequency dependent and are highest near the structural resonance frequency. The changes in the sound pressure level due to structural resonances are audible as amplitude modulations in the vehicle interior and are therefore considered for the generation of the stimuli of the preliminary experiments.

Gearbox

The gearbox is the most dominant tonal sound source of an electric vehicle. Even though many pure-electric vehicles do not have a switchable gearbox, the gearbox contributes largely to the overall vehicle interior noise for both pure-electric and hybrid vehicles. While for vehicles with combustion engines, the level is typically between 10 and 20 dB SPL below the level of the combustion engine [6], in the interior of electric vehicles, the intake from the gearbox is rather prominent. Especially during acceleration and coast-down driving conditions, the planetary gears emit distinct tonal components, which are related to the underlying rotational rate of the electric motor. In the case of a single-speed gearbox, as in many pure-electric vehicles, the frequencies of the emitted tones are approximately proportional to the vehicle speed. As the frequencies of the emitted tones are commonly located in the frequency range where the human auditory system is most sensitive, the sound emissions from the gearbox could severely decrease customer satisfaction. However, they provide acoustic feedback to the driver about the current driving condition [3].

The main type of noise originating from the gearbox is the gear whining (also often called gear whistling), which is generated by different generation mechanisms: The so-called parameter-excited oscillations result from the elasticity of the gear wheel teeth and due to productional and operational inaccuracies. Another noise generation mechanism is the periodic impacts during gear meshing, which are caused by deviations from the ideal gear meshing path resulting from elasticity differences within the gear teeth [6]. The third noise generation mechanism is the rolling noise resulting from friction during teeth meshing, which mainly depends on the surface roughness of the gear wheels [7]. Similarly, to the electric motor orders, the emitted tones are commonly referred to as gear orders because the tone frequencies are directly related to the rate of rotation of the electric motor. From observations, the frequency of the gear orders is commonly lower than of the electric motor orders, so the intake into the vehicle interior is larger than of the electric motor orders. Furthermore, the auditory sensitivity in the mid-frequency range is higher than for the higher-frequency sounds originating from the electric motor [8].

The other noise type from the gearbox is gear rattling, which is generated by the backlash of auxiliary parts of the gearbox. In contrast to the gear whining, the gear rattling has a minor relevance in the vehicle interior and can therefore be neglected during the interior sound analysis [6].

Inverter

The inverter converts the direct current from the battery to alternating current to supply the electric engine. Therefore, a pulse width modulation (PWM) approach is used to generate an approximately sinusoidal voltage. The acoustic emissions of the inverter mainly depend on the inverter switching frequency [4]. A higher switching frequency leads to a smoother approximation of the resulting sinusoid, resulting in a lower noise excitation. As a drawback, the electric losses due to switching are higher for higher switching frequencies [9]. Common switching frequencies are found to be between 8 and 10 kHz [4, 9]. In addition to the switching frequency, also multiples and partials could be visible in the spectrogram.

Besides the components directly resulting from the switching frequency, interactions between the switching frequency and higher harmonics of the electric motor characterize the noise emissions from the inverter [4, 10], which are commonly described as inverter harmonics. As the interactions between the switching frequency and higher harmonics from the electric motor depend on the rate of rotation of the electric motor, the inverter components in the spectrogram commonly appear umbrella-shaped (see Figure 1). Even though the high-frequency tonal character of the inverter components could affect the pleasantness severely, own observations revealed that they are rarely audible in the vehicle interior due to their low sound pressure level.

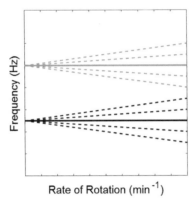

Rate of Rotation (min^{-1})

Figure 1. Schematic illustration of the inverter switching frequency (solid blue line) and the respective inverter harmonics (dashed blue lines). The yellow solid line represents a multiple of the switching frequency and the yellow dashed lines the respective inverter harmonics.

Broadband Tire-Road Noise

The dominant contributor at lower vehicle speeds to the overall loudness in the vehicle interior is the tire-road noise, which results from the contact between the tire surface and the road. In contrast to other sound components, they give useful feedback about the road surface conditions to the driver [6]. The tire-road noise results from a large number of different physical principles. Four main noise generation principles are distinguished [11, 12]:

- Block harmonics
- Air pumping through the tread pattern
- Tire-cavity vibrations
- Tube resonances

The block harmonics have the largest contribution to the overall tire-road noise [13]. They result from the excitation of the tread blocks due to contact with the road surface. To avoid a tonal character of the noise component resulting from the block harmonics they commonly do not have the same length. The air pumping through the tread pattern has the second largest contribution to the overall tire-road noise. It is generated by periodical openings and closures of cavities within the tread pattern. Due to the weight of the vehicle, the air is compressed in the cavities of the tread pattern. The sudden release of the air during the opening of the cavities generates a high-frequency noise, whose frequency is commonly higher than 1000 Hz. Because the frequency of the tire-cavity vibrations mainly depends on the radius of the tire (with a little influence of the vehicle speed), they elicit an almost constant tone. The tire-cavity vibrations have a minor contribution to the overall tire-road noise. They are a typical low-frequency component and result from the excitation of the air inside the tire. The tube resonances are generated by longitudinal and transversal grooves of the tread [11]. As the frequencies of the tube resonances only depend on the length of the tire-to-road contact, the emitted noise is also mostly independent of the driving condition. All the previously mentioned noise components resulting from the described generation mechanisms are amplified by the so-called horn effect: The tread of the tire and the road surface form a convex resonance cavity. This cavity, which is commonly described as a horn, could amplify the tire-road noise by up to 20 dB. Therefore, the intake of the tire-road noise into the vehicle interior is a major contributor to the total vehicle interior noise, which is independent of the propulsion method of the car [11, 14, 15, 16].

Wind Noise

Another major interior noise source of conventional vehicles with combustion engines as well as of vehicles with electrified drives is the sound resulting from the airstream of the vehicle in motion. Especially at higher driving speeds, wind noise becomes the dominant noise source in the vehicle interior and contributes the largest portion of the overall loudness [3].

For conventional vehicles, the intake from the wind noise becomes the dominant noise source at vehicle speeds between 110 [17] and 120 km/h [18], while for electric vehicles already for vehicle speeds below 100 km/h the wind noise will be the primary contributor to the overall vehicle sound due to the absence of an internal combustion engine [3]. For the primary wind noise generation, three different formation mechanisms are distinguished [11]:

- Edge noise generated by leading and trailing edges
- Whistle noise
- Cavity resonances

Edge noises primarily result from turbulent vortex shedding at the leading and trailing edges of the airstream. Edge noise commonly originates from the wheel wells, the trailing edge of the vehicle or the underbody [11, 17]. Especially turbulences from the underbody could excite the floor pan. The resulting rumbling in the vehicle interior [11] strongly decreases the pleasantness of the interior sound [19]. Whistle noise is considered a special kind of trailing edge noise. Due to certain geometric properties, turbulences cause periodic fluctuations resulting in audible high-frequency tonal components. Common noise sources for high-frequency whistle noise are hang-on parts like the exterior mirrors or joints between different parts of the vehicle body. Cavity resonances result from periodic pressure fluctuations inside cavities, e.g. from open windows or sunroofs [20]. Cavity resonances could further cause audible modulations, which are perceived as rumbling [19]. The absence of a combustion engine considerably changes the perception in the vehicle interior for lower speeds due to the unfamiliar dominance of wind noise in those speed ranges while for higher vehicle speeds the interior noise of an electrified vehicle is similar to the one of a conventional vehicle with a combustion engine. The addition of the wind noise and the previously described tire-road noise results in a background noise with a low-pass shape, which increases with an increase in the vehicle speed.

Auxiliary Noise

A major difference to vehicles with an internal combustion engine is that during idle, the electric engine is not running and does not emit any noise. Due to that silence, auxiliary noises, such as air conditioning, steering pump and fan will be considerably more prominent [3]. Even though they could severely affect the pleasantness of the sound in the vehicle interior, especially at low driving speeds, the conduction of recordings under controlled conditions is rather challenging. Therefore, auxiliary noises are not within the scope of this work.

Internal Combustion Engine Powertrain

The present work describes the automatic assessment of the pleasantness of vehicles with electrified powertrains. As mentioned in the Introduction, this category does not only include pure-electric vehicles, but also hybrid vehicles. The separation, allocation and assessment of vehicles with combustion powertrains were conducted in several previous studies [21, 22, 23] and are not within the scope of this work. However, they still influence the perception of hybrid vehicle interior sounds. Therefore, a short overview of the combustion-engine-related noise components is provided.

The most obvious noise originating from the powertrain is the combustion noise, which primarily depends on the parameters rate of rotation and torque. The combustion noise could be further separated into the noise emissions from the internal combustion engine, which are mainly transferred as structure-borne noise through the bearing components and the gas exchange noise, which is mainly transferred as airborne noise through the exhaust system [11]. Especially at low vehicle speeds, the combustion noise dominates the interior sound of hybrid vehicles, so that low-frequency electric motor and gear orders are partially or completely masked. The tonal perception is therefore reduced, while the level at that speed is considerably higher than that of the same vehicle driven in the electric mode. At higher vehicle speeds, similar to conventional vehicles with a combustion engine, the tire-road and wind noise components dominate the interior noise, so that at higher vehicle speeds, the perception of a hybrid car in the vehicle interior does not differ much from the one of a pure-electric car.

For vehicles with a diesel engine, the diesel knocking strongly influences the perception of the interior sound. It results from the high pressure gradient during the combustion of the air-fuel mixture and elicits a distinct impulsive noise reducing the pleasantness of the interior sound [24]. Another impulsive noise is the ticking noise, mainly originating from the fuel injectors. Even though the sound pressure level is considerably lower than the noise from the combustion engine, the increase of the sharpness could negatively affect the pleasantness, especially at low rates of rotation and low engine loads where the sound pressure level of the combustion noise is low [25].

Conventional drivelines further emit tonal components, e.g. the gear orders and the whining sound from the turbocharger. The generation of gear orders within conventional vehicles and hybrid vehicles is similar to the generation within electric vehicles. Modern combustion powertrains use a turbocharger to compress the combustion air, which increases the efficiency of the combustion engine. As a common drawback, turbochargers emit various sounds, including whining, hissing and pulsation sounds.

Doleschal et al., 2020 [26] showed that similarly to the approach described in section 3.5, the audibility of separated sound components for combustion powertrains could be carried out using an auditory masking model [27]. On that basis, the influence of the masked threshold could be assessed when conducting experiments on the variation of specific sound components.

2.2 Sound Separation and Classification Algorithm

The electric vehicle interior noise consists of various noise components, such as the orders from the electric motor and the gearbox, the inverter components, the tire-road noise and the wind noise. To evaluate the influence of single components on the perception of the vehicle interior sound, it is useful to separate the recorded interior noise into its components. Therefore, the separation algorithm according to Fröhlingsdorf and Pischinger, 2022 [2] was used. The resulting separately listenable sound components have been used throughout the dissertation for various purposes: The most important purpose was to enlarge the set of stimuli using a data augmentation approach. The separated sound components enable the amplification, attenuation and removal of sound components of a recorded sound so that the influences of single components on the perceived pleasantness and magnitude of tonal content could be evaluated. These investigated influences could be used in a later step for the implementation of active and passive measures to improve the pleasantness of the interior sound. One of these measures could be the generation of partial sounds (subharmonics) using a vocoder-based approach, which enables to lower the perceived pitch of audible tonal components. A further use-case for the separated sound components is the estimation of the masked threshold, which was conducted experimentally, by running a modified version of an auditory masked model and by using a rough estimation, the so-called critical masking ratio: The masked threshold is the minimum required level at which a certain sound component ("target") is audible assuming a certain background noise ("masker"), so that components, whose actual level is below the masked threshold, are totally masked and do not have to be considered for further analysis.

Electric Motor and Gearbox

The separation of tonal components was conducted using image-processing algorithms developed by Fröhlingsdorf and Pischinger, 2022 [2]. Within the separation process, it was further distinguished between the separation of the electric motor and gear orders and the sound components originating from the inverter.

In addition to the input sound, the separation algorithm further uses a rate of rotation input signal to calculate the order spectrogram [28]. The order spectrogram is scaled in a way that the abscissa is the time axis and the ordinate displays the frequency as multiples of the rotational rate, the so-called orders. The order spectrogram is then normalized and transformed into a binary image using a fixed threshold value. While a lower threshold value increases the number of detected noise artefacts, a higher noise level hampers the detection of electric motor and gear orders, so both issues had to be weighed against each other to set the optimal threshold value. A subsequent line detection algorithm using the Hough transformation [29] was used to detect lines within the normalized and binarized image. The resulting lines either could originate from the electric motor, the gearbox or could be mistakenly detected noise artefacts.

To allocate the lines to the respective sound sources (electric motor or gearbox), a linear support-vector machine (SVM) algorithm is used. The main characteristic of orders resulting from the electric motor is that they involve stronger amplitude modulations resulting from structural resonances. Furthermore, the order number is commonly higher than for orders resulting from the gearbox. Therefore, both the input parameters variance of amplitude and order number were used as input parameters of the SVM. To separate the remaining gear orders from noise artefacts, a weighted nearest neighbor algorithm [30] was used. In addition to the two input parameters order number and variance of amplitude, which are used for the separation of the electric motor orders, the mean value of the positive amplitudes is used. Because the amplitudes of noise artefacts are commonly much lower than those of gear orders, the amplitude is a suitable classification criterion. Altogether, the separation algorithm for the electric motor and gear orders can detect 97 percent of the electric machine and gear orders. From these 97 percent, 98.6 percent of the electric motor orders were correctly identified as electric motor orders and 99.4 percent of the gear orders were correctly identified as gear orders.

Inverter Components

The sound components emitted by the inverter could be separated into two different types: The switching frequency including their multiples and partials is independent of the driving condition. In contrast, the inverter harmonics result from interactions between the inverter and higher orders of the electric motor. They diverge from the switching frequency in an umbrella-shaped way with an increasing rate of rotation of the electric motor. Therefore, the separation of both inverter component types was conducted in two different ways:

In contrast to the electric motor and gear orders, the switching frequency components are independent of the rate of rotation, so a time-frequency spectrogram instead of an order spectrogram was calculated and normalized. The detection of the horizontal inverter switching frequency lines was conducted using a Kirsch-Compass filter, which detects lines by calculating the gradient in each direction and choosing the highest gradient [31]. The binarization of the image was conducted according to the algorithm of Otsu, 1979 [32], which is a nonparametric and unsupervised method for the selection of an optimal discrimination threshold. The previously described Hough line detection [29] enables the detection of the switching frequencies in the spectrogram.

The second type of inverter component is the inverter harmonics. As they result from interactions between the inverter and higher harmonics of the electric motor, the detection is conducted based on previously detected inverter switching frequencies and their multiples. In some cases, inverter harmonics of invisible switching frequencies are visible in the spectrogram. Therefore, also possible multiples and partials of the detected switching frequencies are considered. The detection of the inverter harmonics is based on a fixed-value binary spectrogram similar to the electric motor and gear orders. The angles of the inverter harmonics to the respective switching frequency depend on the number of pole pairs of the electric motor, so that the geometric data of the electric motor, if provided, could be used for an initial guess for the detection of harmonics during the following Hough line detection. In total, the algorithms detected 94 percent of the inverter switching frequencies and 86 percent of the inverter harmonics, which enables the usage of the separation algorithm for the data augmentation experiment described in section 6.2.

Tire-Road and Wind Noise

The separation of tire-road and wind noise components was conducted after removing the components originating from the electric motor, the gearbox and the inverter. Within the tire-road noise components, a distinction has to be made between the tire-cavity vibrations, which are visible as tonal components in the spectrogram and have a similar character as electric motor orders, and the residual noise components originating from the other generation mechanisms. The tire-cavity vibrations were separated in a similar process to the inverter switching frequencies: A Kirsch-Compass filter filters the normalized spectrogram, which is followed by an Otsu binarization and a Hough line detection algorithm. The algorithm described above is able to detect 91 percent of the tire-cavity vibrations. The tire-cavity vibrations are different from the electric motor orders in a way that the masking by the other tire-road and wind noise components is considerably stronger due to the lower frequency range of the tire-cavity vibrations and the low-pass character of the noise. Furthermore, the frequency of the tire-cavity vibrations mainly depends on the diameter of the tires and only to a lesser extent on the current vehicle speed, so the frequency of the tire-cavity orders is almost constant over time. Due to the similarity of the acoustic properties to those of the electric motor and gear orders, a separate experiment on the variations of tire-cavity vibrations has not been conducted. Instead, it is assumed, that the implications are the same as for the electric motor orders.

The residual noise is further separated into the tire-road and wind noise components. Therefore, an operational transfer path analysis (OTPA) [33] was conducted to calculate the transfer functions for both types of noise.

Observations indicated that the relation between the gradient of the level and the velocity of the vehicle is similar for all investigated vehicles for both the tire-road and wind noise components. In contrast, the overall level of the tire-road noise depends much on the tire type: In general, the emissions from winter and all-season tires are considerably higher than from those of summer tires. The wind noise depends on the size of the vehicle: The intake of wind noise into the vehicle interior is considerably higher for small and compact cars than for larger vehicles. Therefore, the median of the transfer functions resulting from the OTPA was taken to separate the residual noise into the tire-road and wind noise components.

2.3 Pleasantness and Magnitude of Tonal Content

Several studies investigated the pleasantness, annoyance or sound quality perception: Fastl and Zwicker, 2007 [8] developed a general model for all kinds of sound to estimate the relative pleasantness of one sound to another sound (see chapter 9.4 of the book). The formula contains the parameters roughness, sharpness, tonality and loudness. They further developed another formula to estimate the psychoacoustic annoyance, containing the parameters N_5 percentile loudness, sharpness, fluctuation strength and roughness [8] (see chapter 16.1.4 of the book). It should be noted that several studies [34, 35, 36] indicate that there is a strong relationship between the attributes of pleasantness and annoyance even though they are no exact opposites. The attribute pleasantness (German term "Angenehmheit") is better suited for the assessment of vehicle interior sounds because the attribute annoyance implies exposure to unwanted sounds [35]. Vehicle interior sounds are not necessarily unwanted. Interviews with listeners, who participated in the hearing experiments, which were conducted during the current study, revealed that vehicle interior sounds could be pleasant and that the pleasantness of the vehicle interior sound is a major criterion when buying a vehicle. The terms pleasantness and annoyance have to be further distinguished from the term "sound quality", as this term refers to the extent to which sound properties meet certain requirements, which have to be defined in advance [37]. Therefore, the pleasantness is better suited for the experimental evaluation and the subsequent model development.

Apart from general models, several studies have been conducted to develop specific models for the assessment of vehicle sounds. Murata et al.,

1993 [38] investigated the relation of several attributes of a semantic differential experiment on the perceived sound quality. Conducting the following factor analysis, they resumed that a combination of the factors comfort, power and booming (as an opposite of ringing) is a good explanatory model for the perceived sound quality. Schumann et al., 2019 [21] developed a log-linear equation to estimate the pleasantness of interior sounds of vehicles with combustion engines depending on the psychoacoustic parameters loudness, sharpness and fluctuation strength. Equation-based regression approaches have the common drawback that a certain functional relation has to be assumed in advance. Therefore, Ma et al., 2017 [39] used a feedforward neural network to automatically assess the pleasantness of pure-electric vehicle interior sounds, considering the parameters A-weighted sound pressure level, loudness, fluctuation strength, tonality, roughness, articulation index and sharpness. A following calculation of the weight coefficients according to the algorithm of Garson, 1991 [40] showed, that the parameter A-weighted level influenced the sound quality at most, even though the weights of the other coefficients were not much lower. Steinbach and Altinsoy, 2019 [41] used a similar approach as Ma et al., 2017 [39] showing that also for pass-by sounds of electric vehicles, the A-weighted sound pressure level has the strongest influence on the perceived annoyance, which was further confirmed by a subsequent sensitivity analysis.

The MOTC describes the magnitude of the audibility of tonal components within a specific background noise. The perceived MOTC depends on the level and the spectrum of both the tonal component and the background noise. Several studies postulate that, in general, the perceived MOTC negatively influences the pleasantness of electric vehicle interior sounds [42, 43, 44]. Therefore, the commonly used standards to assess the MOTC, such as DIN 45681 [45] and ISO 1996-2 [46] assume a negative relation between the two sensations and propose tone adjustments, the addition of up to 6 dB to consider the prominence of audible tones.

However, the addition of subharmonics could increase the perceived pleasantness [47, 48, 49], even though the MOTC increases due to the higher number of tones [49]. Therefore, the previously postulated assumption is not always useful for the assessment of electric vehicle interior sounds, as more and more vehicles use active sound design techniques in the vehicle interior sound, where subharmonics are an essential design element [50]. Thus, a value-free rating scale for the audibility of separate tones should be used, such as the ECMA-418-2 standard (formerly known as ECMA-74) [51]: It uses a linear scale for the perceived MOTC. The scale is calibrated using a pure tone with a frequency of 1 kHz and a sound pressure level of 40 dB SPL, whose tonality is defined as one tu$_{HMS}$ (tonality units according to the Hearing Model of Sottek). In contrast to the DIN 45681 standard and the Pure Tonalness according to Parncutt, 2012 [52], the ECMA-418-2 standard is further suitable to assess the time-varying MOTC [53], because tonal components of electric vehicle interior sounds are primarily audible during transient driving conditions [3]. Therefore, the MOTC according to the ECMA-418-2 standard was used for the pleasantness model development.

2.4 Active Sound Design

Many approaches and techniques have been carried out to improve the interior sound of both electric and internal combustion engine vehicles, which include both passive and active measures. The passive measures could be further separated into the reduction of the noise emissions from the emitting sources, such as the reduction of electromagnetic noise, stiffening of the drive unit, its mount design and adjustments of the structural transfer path [50] and insulation measures to block off sounds so that they cannot be transferred into the vehicle interior [54].

The active sound design describes measures to insert artificial sound components, which could be both tonal components and noise. Bodden and Belschner, 2016 [55] formulated key requirements for components, which are inserted into the interior sound of electric vehicles as a part of an active sound design concept:

- Authenticity
- Long-term applicability
- Discriminability from combustion sounds

The first criterion authenticity describes that the inserted sounds should sound like they originate from the vehicle, which implies that the current driving condition has to be considered for the generation of the sound components. The second criterion is long-term applicability: The artificial sounds should therefore not annoy the vehicle passengers if they are applied for long durations. Furthermore, they should clearly be abstracted

2 State of Research

from combustion engine sounds. Moon et al., 2020 [56] stated that active sound measures could not only be used to improve the pleasantness of an unpleasant interior sound but further offer an opportunity to represent a certain brand identity in the vehicle interior. Additionally to the brand identity, the sound can be customized to the vehicle passengers and changed during the lifetime of the vehicle, because the generation is detached from the mechanical processes within the vehicle. The implementation of active measures could further save costs compared to adjustments of the powertrain mechanics [57].

Different application techniques of active sound design measures are distinguished. One possible approach is to passively generate the sound components by directly using the structure-borne noise emitted by certain components and to create a transfer path into the vehicle interior to make the desired components more pronounced. Even though the resulting sound components sound authentic as they originate from real components, the design capabilities are limited since only existing sound components can be used [57]. Another approach is to include a sound generator module into the power electronics of the electric motor so that the electric motor can be used to insert both tonal and noisy sound components into the vehicle interior sound [50]. The third approach is to use the infotainment system of the vehicle and to play back the desired sound using loudspeakers. While this approach offers the highest degree of freedom in terms of the presented sound components, maintaining the authenticity, so that the sound could not be localized from the loudspeakers, could be rather challenging [57].

Different kinds of noise have been proposed to improve the pleasantness of vehicle interior sounds, which could be of both tonal and noisy character. Gwak et al., 2014 [47] and Sun et al., 2018 [48] showed that the application of subharmonics is a suitable approach to increase the pleasantness of the vehicle interior sound by lowering the perceived pitches of the sound. Gwak et al., 2014 [47] suggested the application of noisy components to mask undesired tonal components in future work and stated that they should have a low-pass character to limit the increase of loudness and to avoid an increase of sharpness. He et al., 2021 [50] proposed a combined approach of the application of subharmonics and a so-called dithering noise. Dithering noise describes the technique of generating a time-variant noise with a bandwidth of approximately one critical band-width around undesired components to reduce the salience of the respective pitch and its contribution to the MOTC. This technique is further advantageous in terms that the increase of the parameters loudness (and probably sharpness) is small. A jury experiment revealed for one vehicle, that the combination of the two approaches is suitable to mask tonal whining sounds and to improve the sound quality of the interior sound.

As a consequence of these results, the application of subharmonics and low-pass noise for both synthetic and recorded vehicle interior sounds was evaluated, which includes the implications on both the perceived pleasantness and MOTC.

3 Audibility of Sound Components

Several noise components (e.g. electric motor orders or inverter components) only influence the pleasantness, if they are not masked by their background noise, which consists of the tire-road and wind noise components [39]. Therefore, it is advantageous to determine the sound pressure level, at which the component is just audible within the surrounding noise background. In the following, masked thresholds of sound components are determined experimentally and by means of simulations with two model approaches. One model, the auditory masking model [27], simulates the experimental procedure. Since the threshold estimates with this model are time-consuming due to high computational cost a second model is used for a rough estimate of the threshold of tonal components (but not of noisy components). This model is referred to as the critical masking ratio [58]. The masked threshold estimates of both the auditory masking model and the critical masking ratio are compared to the average masked thresholds of ten normal-hearing listeners. The latter were published in Doleschal et al., 2021 [59].

3.1 Listeners

Ten normal-hearing listeners participated in the experiments for the determination of the masked thresholds. None of them had any history of hearing difficulties, and their audiometric thresholds were 20 dB Hearing Level (HL) or below at the standard audiometric frequencies between 125 and 8000 Hz.

3.2 Apparatus

Listeners were seated in a double-walled sound-attenuating booth. The stimuli were digitally generated in MATLAB, converted to analog signals via the external sound card RME Fireface UC and presented diotically via Sennheiser HD 650 headphones.

3.3 Stimuli

To determine masked thresholds, stimuli with a duration of five seconds were generated. They consisted of a background noise, which simulated tire-road and wind noise components and tonal components, which simulated electric motor and gear orders. The background noise was generated as follows: For each noise sample, a newly generated white noise was filtered with a second-order Butterworth low pass filter with a cut-off frequency of 40 Hz and a slope of 12 dB per octave. Both parameters were adapted from recorded interior sound samples. The masked thresholds of the synthesized 24th and 48th electric motor orders [4] were simulated as sinusoidal sweeps in a run-up (ru), a full-load run-up (fl) and a coast-down (cd) condition. For the experiment, the following conditions were considered:

- Run-up 0-30 km/h
- Full-load run-up 0-60 km/h
- Coast-down 48-12 km/h

The conditions were realized by varying the level of the background noise and the frequency of each tonal component. Every component s_c was realized as a linear sinusoidal frequency sweep according to the equation

$$s_c(t, f_b, f_e) = a * \sin\left(2\pi \int_0^t \left(f_b + \frac{(f_e - f_b)}{T} \cdot \tau\right) d\tau\right) \qquad (1)$$

with the amplitude a, the initial frequency f_b at the beginning and the final frequency f_e at the end of the sound. The total duration of the sound T was always set to five seconds, while the frequencies f_b and f_e were varied according to the driving condition and the simulated order number.

For the ru condition, the frequency ranged from $f_b = 0\,Hz$ to $f_e = 1000\,Hz$ for the 24th order (ru_24) and from $f_b = 0\,Hz$ to $f_e = 2000\,Hz$ for the 48th order (ru_48). Due to the reason, that the speed increase over time in the fl condition was twice as high as of the ru condition, the respective frequency ranges were from $f_b = 0\,Hz$ to $f_e = 2000\,Hz$ for the 24th order (fl_24) and from $f_b = 0\,Hz$ to $f_e = 4000\,Hz$ for the 48th order (fl_48). For the cd condition, where the corresponding vehicle speed decreased from 48 km/h to 12 km/h, the frequency decreased from $f_b = 1600\,Hz$ to $f_e = 400\,Hz$ for the 24th order (cd_24) and from $f_b = 3200\,Hz$ to $f_e = 800\,Hz$ for the 48th order (cd_48) [49, 59].

The speed increase over time for the conditions ru and fl was simulated by a linear increase of the background noise amplitude. The level at the beginning of the sound was 27 dB SPL. It increased up to 77 dB SPL for the *ru* condition and up to 83 dB SPL for the *fl* condition. The initial and final levels were adapted from interior vehicle recordings. The 6-dB difference between the final levels of the *ru* (at 30 km/h) and the *fl* condition (at 60 km/h) is in agreement with the findings of Zeller, 2009 [11]. For the *cd* condition, the amplitude decreased linearly over time, with an initial level of 82 dB SPL and a final level of 72 dB SPL.

3.4 Experimental Procedure

Masked thresholds were measured for both the 24th and the 48th electric motor order of each driving condition (*ru*, *fl* and *cd*) by using an adaptive three-interval three-alternative forced-choice procedure. During each trial, three sounds were presented one after another with 500-ms silence intervals in between. Each sound contained a different realization of the background noise. One randomly chosen interval contained the currently investigated electric motor order of the currently investigated driving condition. The task for the listener was to indicate which of the intervals contained that order ("In welchem Intervall war ein Unterschied zu hören?" ["Which interval is different from the others?"]) by pressing the key "1", "2" or "3" on a number pad.

The initial level at the beginning of the experiment was chosen to be 40 dB SPL because this level was markedly above the threshold for each order of each condition and allowed the listener to become familiar with the target signal. The level was adjusted according to a two-down one-up rule to estimate the 70.7 % point of the psychometric function [60]. If the listener indicated the correct interval two times in a row, the level was decreased by the current step size. After each incorrect answer, the level was increased by the current step size. The initial step size was 8 dB SPL. The step size was halved after each upper reversal (a level increase, followed by a level decrease) until a step size of 1 dB SPL was reached. This step size was kept until six more reversals occurred. The levels at these six reversals were averaged to obtain an estimate of the masked threshold of the investigated order.

Each listener performed three experimental runs for each electric motor order of each driving condition in random order under the constraint, that

each combination of electric motor order and driving condition was selected once before the same combination was reselected. The individual
threshold of each listener for each combination was determined by aver-
aging the threshold estimates of the three runs. The individual thresholds
of all listeners were averaged and the standard deviations were calculated
to compare the results to the predictions of the auditory masking model
and the critical masking ratio.

3.5 Estimation of Masked Threshold

3.5.1 Auditory Masking Model

To simulate the masked thresholds of the above-mentioned electric motor
orders, a modified version of the perception model according to Dau et al.,
1996 [27] was used. The first stage of the model is an outer and middle
ear filter, which was modeled by a band pass filter with cut-off frequencies
of 0.5 and 5.3 kHz [61]. The auditory frequency selectivity was modeled
by a fourth-order gammatone filter bank [62] with a spectral distance of
one ERB_N equivalent (normalized equivalent rectangular bandwidth [63] of
normal-hearing listeners at moderate levels) and a bandwidth of one ERB_N
of each filter. The frequency range of the gammatone filter bank was set
according to the lowest and highest frequency of the synthetic electric motor order. Afterwards, Gaussian white noise was added to the output of
each filter to consider a lower limit for the detection of thresholds in quiet
[64]. The processing in the inner hair cells of the ear was simulated by a
combination of a half-wave rectification and a first-order low-pass filter with
a cut-off frequency of 1 kHz [27]. The adaption and compression processes are simulated by five consecutive nonlinear feedback loops [27]. A
second-order Butterworth low pass filter with a cut-off frequency of 8 Hz
filtered the resulting output [27, 65]. In the following, the output of this process is described as an internal representation.

Afterwards, the experimental alternative-forced choice procedure is simulated. In each trial, the internal representation of each interval is calculated
and compared to a previously calculated representation of a background
noise mixed with the electric motor order at a suprathreshold level. The
observer stage chooses the interval with the highest correlation. The
threshold estimation was simulated according to the process described in
section 3.4.

Due to the reason, that the background noise is transient in level and the electric motor order is transient in frequency, the signal-to-noise ratio substantially changes during the signal duration of five seconds. Because the lowest threshold over time determines the overall masked threshold, the signals were cut into temporal portions of 0.5 s with an overlap of 0.25 s, resulting in 19 portions for each signal. For each temporal portion, the threshold was calculated separately. To check, if the method returns consistent thresholds, the whole process has been run 100 times. The lowest thresholds of each run were averaged and the standard deviations over all lowest thresholds of each run were calculated.

3.5.2 Critical Masking Ratio

The critical masking ratio [58] is a simple and fast method to estimate the audibility of a tonal component. In contrast to the previously described auditory masking model, the critical masking ratio is the ratio of the sound pressure level of the tonal component to the surrounding background noise with a bandwidth of one ERB_N. The basis for the threshold estimation of each tonal component is the tonal component itself and the respective residual noise, which is calculated by subtracting the component from the total sound. The extracted component is usually not purely sinusoidal but has a similar character as a pure tone. Therefore, the component is simplified as a pure tone by calculating the average spectral centroid of the component for each snippet with a duration of half a second. Subsequently, a bandpass filter with a bandwidth of one ERB filters the residual noise. Afterwards, the ratio of the component level to the filtered residual noise level is calculated. If the ratio (in dB) during the snippet is over a certain value, which depends on the frequency of the tonal component, the snippet will be considered audible. The component is considered audible if the tonal component is audible during at least one snippet. Accordingly, the threshold could be calculated by adding the value of the critical masking ratio for each snippet to the background noise level and taking the minimum. Due to the usage of a newly generated noise, the thresholds differ between the different runs of the critical masking ratio calculation. Similarly, to the threshold estimation using the auditory masking model, the calculation was carried out 100 times. Afterwards, the mean values and standard deviations were calculated.

3.6 Results

Figure 2 shows the mean values and standard deviations of the predictions from the auditory masking model and the critical masking ratio in comparison to the individual masked thresholds of the listeners for the 24th and 48th electric motor orders with the respective background noise for the three driving conditions *ru*, *fl* and *cd*. The experimental results show average masked thresholds [interindividual standard deviations in brackets] of 22.5 dB SPL [± 1.1 dB] for the order *ru_24*, 13.9 dB [± 4.1 dB SPL] for *ru_48*, 18.4 dB SPL [± 2.7 dB SPL], for *fl_24*, 8.7 dB SPL [± 2.5 dB SPL] for *fl_48*, 19.3 dB SPL [± 1.2 dB SPL] for *cd_24* and 9.2 dB SPL [± 1.2 dB SPL] for *cd_48*.

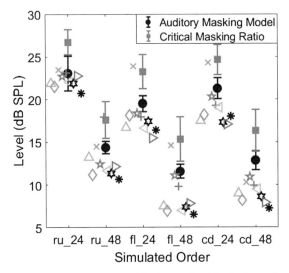

Figure 2. Mean values and interindividual standard deviations for the predictions of the auditory masking model (blue circles) and the critical masking ratio (red squares) in comparison to the individual mean values of the listeners for six different artificial electric motor orders (all other symbols). The mean values and interindividual standard deviations of the experimental results have been previously published in Doleschal et al., 2021 [59].

The simulation by the auditory masking model revealed results of 23.1 dB [±2.0 dB SPL] for the order *ru_24*, 14.4 dB SPL [± 0.7 dB SPL] for *ru_48*, 19.5 dB SPL [± 0.9 dB SPL] for *fl_24*, 11.6 dB SPL [± 0.8 dB SPL] for *fl_48*,

21.3 dB SPL [± 1.2 dB SPL] for *cd_24* and 12.9 dB SPL [± 1.1 dB SPL] for *cd_48*.

The threshold estimates of the critical masking ratio are higher than those of the auditory masking model. The mean estimates were 26.8 dB [± 1.5 dB SPL] for the order *ru_24*, 17.6 dB SPL [± 2.2 dB SPL] for *ru_48*, 23.3 dB SPL [± 2.0 dB SPL] for *fl_24*, 15.4 dB SPL [± 2.6 dB SPL] for *fl_48*, 24.7 dB SPL [± 1.8 dB SPL] for *cd_24* and 16.4 dB SPL [± 2.5 dB SPL] for *cd_48*.

The comparison between the experimental results and the simulations of the auditory masking model shows that, except for the order *cd_48*, the results of the model and the listeners are in good agreement and the standard deviations of the model are considerably lower than the range of the average individual thresholds of the listeners. Averaged over all six electric motor orders, the deviation between the mean value of the model predictions and the average masked threshold of the listeners is 1.8 dB SPL. Furthermore, the average standard deviation of the model predictions was only 1.2 dB SPL. The average calculation times for the six artificial orders ranged from 116 s (*cd_24*) to 175 s (*fl_48*). The long computation times sometimes hamper the practical usage of the model. Therefore, as an alternative, the critical masking ratio was used to give a rough estimation of the audibility of tonal components. Compared to the mean thresholds from the listener, they are between 3.7 (*ru_48*) and 7.2 (*cd_48*) dB SPL higher than the mean values of the listeners. The systematic overestimation of the thresholds indicates that the listeners use additional cues, such as the temporal fine structure of the signal, which are not considered when simplifying the tonal component as a sinusoid. As the critical masking ratio is further unable to simulate the thresholds of noisy components, the more time-consuming auditory masking model should be generally preferred.

4 Experiments with Synthetic Sounds

In this chapter, psychoacoustic experiments are described that measure pleasantness and MOTC. These experiments use artificial synthetic sounds that contain the main characteristics of recordings of the interior sound of electric vehicles. In addition, augmented versions were tested to investigate the influence of parametric signal variations on both the pleasantness and the MOTC. Parts of the experimental results have been previously published in Doleschal et al., 2021 [59] and Doleschal and Verhey, 2022 [49].

4.1 Setup and Methodology

Twenty normal-hearing listeners participated in each experiment for both the variables pleasantness and MOTC. None of them had any history of hearing difficulties and their audiometric thresholds were 20 dB HL or below at the standard audiometric frequencies between 125 and 8000 Hz.

Prior to each experiment, all sounds of the experiment were presented to the listener to familiarize the listener with the general characteristics of each sound and the variability of both investigated variables. After the introduction, each listener rated the pleasantness of each sound in a magnitude estimation experiment, in some cases followed by a magnitude estimation experiment regarding the MOTC. In some of the experiments, pairwise comparison experiments for the pleasantness and/or the MOTC were conducted additionally for the whole dataset or a subset to validate the results of the magnitude estimation experiment and to quantify the magnitude of both sensations on a ratio scale.

For the pleasantness rating, each sound of the respective experiment was rated on a scale between the extreme categories "nicht angenehm" ("not pleasant") and "extrem angenehm" ("extremely pleasant'). The ratings were converted into values from 0 to 100, where 0 indicates the category "not pleasant" and 100 the category "extremely pleasant". Further supportive categories were provided (English translation and corresponding scale value in parentheses): "sehr wenig angenehm" ("very little pleasant", 10), "wenig angenehm" ("little pleasant", 30), "mittelangenehm" ("moderately pleasant", 50), "angenehm" ("pleasant", 70) and "sehr angenehm" ("very pleasant", 90). The same scaling was also used by Doleschal et al., 2021

[59] and Doleschal and Verhey, 2022 [49], which is adapted from the method of categorical loudness scaling [66]. For the magnitude estimation experiment of the MOTC, the same scale was used with the difference that the adjective "angenehm" ("pleasant") was replaced by "tonhaltig" ("tonal"). Each listener conducted three experimental runs for both variables pleasantness and MOTC. For each sensation, three experimental runs were carried out. Afterwards, the mean values and standard deviations over all listeners were calculated.

In each pairwise comparison experiment, every sound was compared to all other sounds of the selected subset. The number of comparisons could be calculated by the binomial coefficient

$$\binom{n}{2} = \frac{n!}{2 * (n - 2)!} \tag{2}$$

with n as the number of elements of the selected subset. In each trial, two sounds were presented consecutively separated by an inter-stimulus silence interval of half a second. The listeners had to indicate, which of the two sounds was more pleasant ("Welches Signal war angenehmer?") or more tonal ("Welches Signal war tonhaltiger?"). The sound pairs for each trial were selected according to the algorithm of Ross, 1934 [67] to avoid regular repetitions and to maximize the spacing between each repetition of the same sound. Every listener conducted two runs for each pairwise comparison subset. After data collection, a ratio scale according to Bradley, Terry and Luce (BTL) [68, 69] was fitted to the data by using an open-source MATLAB function [70]. The resulting BTL scaled values were compared to the magnitude estimation results to further quantify perceptual differences based on a ratio scale.

For all experiments, the listeners were seated in a double-walled sound-attenuating booth. The stimuli were digitally generated using MATLAB, converted from digital to analog signals with the sound card RME Fireface UC and presented via Sennheiser HD 650 headphones. For the preliminary experiments with synthetic sounds, the sounds were presented diotically. For the experiments with recorded and augmented recorded sounds, two-channel recordings were used, which were recorded at the left and right ears' positions of the driver's seat. All signals were switched on and off with 100-ms raised-cosine ramps.

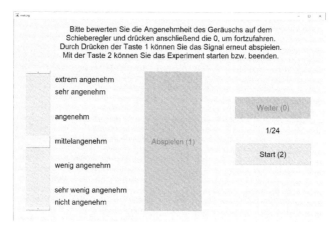

Figure 3. Experimental screen, which was used to conduct the magnitude estimation experiment for the variable pleasantness. For the variable magnitude of tonal content (MOTC), the adjective "angenehm" ("pleasant") was replaced by "tonhaltig" ("tonal").

For the magnitude estimation experiment, the listeners' task was to rate each sound by moving a stepless mechanical slider on a self-built device. The position of the mechanical slider was coupled to the slider visualization on the experimental screen (see Figure 3). The listeners were allowed to listen to the sounds multiple times and to adjust their rating before they proceeded with the next sound. For the pairwise comparison experiment, during each trial, the screen highlighted the currently playing interval and it was not possible to listen to the sounds repeatedly. The listeners indicated the more pleasant/tonal interval by pressing the key "1" or "2" on a number pad.

4.2 Synthetic Electric Vehicle Interior Sounds

4.2.1 Stimuli

In the first experiment, artificial sounds with a duration of five seconds were generated, which contain the key elements of interior recordings. The first part of the sound name indicates the driving condition (*ru*, *fl* or *cd*). The second part of the sound name indicates whether the sound contained none, one or two electric motor orders: For each driving condition, one reference sound did not contain any order (*_ref*). Half of the remaining sounds only contained the 24th (*_1*) and the other half both the 24th and

31

the 48th electric motor orders (_2). They were realized as linear frequency sweeps according to equation (1) and section 3.3. They further contained the same kind of noise for the three conditions, which are described in that section. The level of the 24th order is the third part of the sound name and was set to either 25 (_25) or 40 dB SPL (_40). The choice for the values was motivated by interior recordings, where a level of 25 dB SPL represented a typical case in the vehicle interior (see also Figure 10); while a level of 40 dB SPL represents an extreme case, which usually does not occur in series vehicles. If the sounds additionally contained a 48th electric motor order, its level was always 6 dB lower than the level of the 24th order.

Most of the vehicle interior sounds involve amplifications at certain points in time, which result from structural resonances within the vehicle [71, 72]. Structural resonances commonly increase the MOTC at these points in time [73] and could therefore affect the pleasantness. These structural resonances commonly occur at certain frequencies in the vehicle. Observations from interior recordings indicated that common frequency ranges for structural resonances are approximately 500-560 Hz, 680-740 Hz and 800-820 Hz for part-load run-up conditions. For full-load run-up conditions, the frequency ranges are the same, but the amplifications are stronger than for the part-load run-up conditions. For coast-down conditions with recuperation, the common frequency ranges are 440-540 Hz and 640-680 Hz. The structural resonances were realized using a peaking equalizer filter [74], which enables precise amplification at the resonance frequencies. The peaking equalizer filter is defined by the parameters center frequency f_c, the maximum amplification A and the quality factor Q, which determines the edge steepness of the filter curve. These parameters were adjusted in a way that the cutoff frequencies of the filters are the lower and upper frequencies of the frequency ranges mentioned above and that the resulting amplifications sound realistic when being compared to electric vehicle interior recordings (see Table 1). If a sound contained both orders, the same structural resonances were applied to the 24th and 48th electric motor orders.

Table 1. Filter parameters for the peaking equalizer filters used to simulate the structural resonances.

Driving Condition	Center Frequency f_c	Max. Amplification A	Quality Factor Q
ru	530 Hz	15 dB	3.8
ru	710 Hz	15 dB	5.1
ru	810 Hz	15 dB	17.5
fl	530 Hz	18 dB	3.2
fl	710 Hz	18 dB	4.2
fl	810 Hz	18 dB	14.5
cd	490 Hz	15 dB	2.1
cd	660 Hz	15 dB	7.2

For every driving condition, all possible combinations of structural resonances, including no resonances (_none_), were realized, resulting in eight combinations for the conditions ru and fl and in four combinations for the condition cd. The last part of the sound name contains the center frequencies of the structural resonances, which were added to the sound. As an example, the sound ru_2_25_530_710_880 contains the 24th and 48th electric motor orders and the noise of the run-up condition. The 24th electric motor order has a level of 25 dB SPL, and the 48th order has a level of 19 dB SPL. Both orders have structural resonances, where the level at the center frequencies 530 Hz, 710 Hz and 810 Hz increases by 15 dB, which leads to a maximum level of 40 dB SPL for the 24th order and 34 dB for the 48th order.

Combining all parameter variations results in 33 sounds for the ru and fl conditions and 17 sounds for the cd condition.

4.2.2 Results

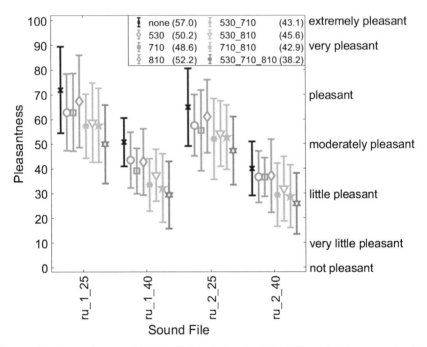

Figure 4. Mean values and interindividual standard deviations for the perceived pleasantness of the run-up (ru) condition. The labels on the abscissa indicate whether the 48th electric motor order was present (_2) or not (_1) and the level of the 24th order. Colors and symbols indicate the number and center frequencies of the simulated structural resonances. The values in parentheses refer to the mean values of all stimuli, which contain the designated structural resonances. Data representation as in Doleschal et al., 2021 [59].

Figures 4, 5 and 6 show the mean values and standard deviations for the pleasantness of the driving conditions *ru*, *fl* and *cd*. The data are grouped in sets of eight (*ru* and *fl*) and four (*cd*) results with the same combination of absence/presence of the 48th order and 24th order level. Note that due to the goal to point out the influence parameters on the perceived pleasantness, the pleasantness of the _ref stimuli is not shown in the figures.

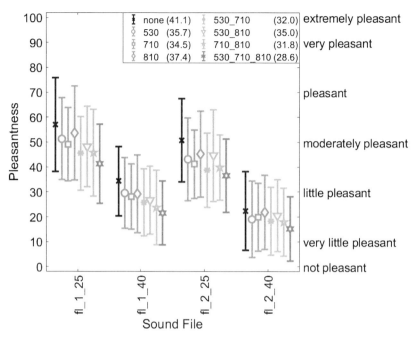

Figure 5. Same as Figure 4, but here for the fl condition. Data representation as in Doleschal et al., 2021 [59].

Average pleasantness ratings range from 26.0 (*ru_2_40_530_710_810*) to 73.7 (*ru_ref*) for the *ru* condition, from 15.1 (*fl_2_40_530_710_810*) to 66.2 (*fl_ref*) for the *fl* condition and from 26.6 (*cd_2_40_490_660*) to 68.1 (*cd_ref*) for the *cd* condition. To indicate which factors influence the pleasantness, a three-way analysis of variance (ANOVA) with a significance level of p=0.05 with the dependent variable pleasantness and the independent variables presence of 48th order, level of 24th order and combination of structural resonances was carried out. For the statistical analysis, the _ref sounds were not taken into account. As the parameters of the structural resonances were different for the three driving conditions, the ANOVA was calculated separately for the three driving conditions.

35

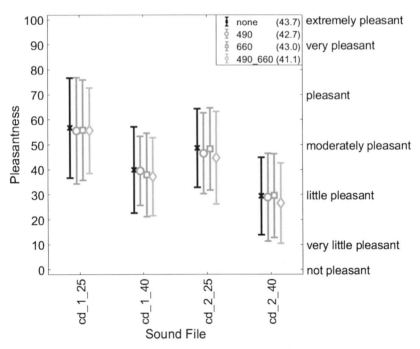

Figure 6. Same as Figure 4, but here for the cd condition. Data representation as in Doleschal et al., 2021 [59].

For the *ru* sounds, the ANOVA revealed, that all of the aforementioned independent variables have a significant effect on the perceived pleasantness, while none of the interactions appeared to be significant (see Table 2). The results showed that when the sounds contained a 48th order (_2) the pleasantness was lower than for the sounds without a 48th order (_1). On average, the pleasantness reduced from 49.9 to 44.6. When the level of the 24th order was 40 dB SPL (_40), the average pleasantness was lower than for the sounds with a 24th-order level of 25 dB SPL (_25). On average, the pleasantness reduced from 58.4 to 36.1. The ANOVA further showed, that for the *ru* condition, the occurrence of structural resonances influences the perceived pleasantness.

To show, which structural resonances significantly alter the perceived pleasantness, a Tukey post-hoc test was carried out. The test revealed that, compared to the situation without subharmonics, both the appearance of the structural resonances _530 (-6.77, 95%-confidence interval (CI) [-13.35, -0.2], p=0.038) and _710 (-8.39, 95%-CI [-14.96 -1.82], p=0.003) caused a significant reduction of the pleasantness as well as the combination _530_710 (-13.87, 95%-CI [-20.44, -7.3], p<0.001).

Table 2. Results of a three-way ANOVA regarding the dependent variable pleas-antness. The independent variables were presence of 48th order, level of 24th order, combination of structural resonances and their interactions. The ru con-dition is shown here. Non-significant variables and interactions were omitted in this table.

Factor	Sum of Squares	Degrees of Freedom	Mean Square	F	p
Presence of 48th Order	4528.54	1	4528.54	24.24	<0.001
Level of 24th Order	80243.3	1	80243.3	429.46	<0.001
Combination of Structural Resonances	19992.48	7	2856.07	15.29	<0.001

In contrast, the occurrence of the resonance _810 only had a significant influence on the pleasantness compared to the situation without in combi-nation with at least one other resonance: _530_810 vs. _none (-11.36, 95%-CI [-17.94, -4.79], p<0.001), _710_810 vs. _none (-14.07, 95%-CI [-20.65, -7.5], p<0.001) and _530_710_810 vs. _none (-18.77, 95%-CI [-25.34, -12.2], p<0.001). Between the other combinations of structural res-onances, further significant differences were found: _530 vs. _530_710 (-7.1, 95%-CI [-13.67, -0.52], p=0.024), _530 vs. _710_810 (-7.3, 95%-CI [-13.87, -0.73], p=0.018), _530 vs _530_710_810 (-12, 95%-CI [-18.67, -5.42], p<0.001), _710 vs. _530_710_810 (-10.38, 95%-CI [-16.95, -3.8], p<0.001), _810 vs. _530_710 (-9.07, 95%-CI [-15.64, -2.5], p=0.001), _810 vs. _710_810 (-9.28, 95%-CI [-15.85, -2.7], p=0.001), _810 vs. _530_710_810 (-13.97, 95%-CI [-20.54, -7.4], p<0.001), _530_810 vs. _530_710_810 (-7.41, 95%-CI [-13.98, -0.83], p=0.015).

Similar results appeared from the results for the fl condition, where all in-dependent variables significantly influenced the pleasantness, while their interactions did not (see Table 3). On average, the pleasantness of the sounds with a 48th order was lower (30.9) than without (38.1). In addition, a 24th order level of 40 dB SPL instead of 25 dB SPL resulted in a signifi-cant pleasantness decrease from 45.7 to 23.3. The appearance of structural resonances further significantly reduced the pleasantness. Therefore, also for the *fl* condition, a Tukey post-hoc test was carried out. In contrast to the *ru* condition, for the *fl* condition, the appearance of only one structural resonance did not significantly reduce the perceived pleasantness. Only the appearance of the combinations of structural resonances *530_710* (-9.1, 95%-CI [-16.42, -1.75], p=0.004), *710_810* (-9.34, 95%-CI [-16.68, -2], p=0.003) and *530_710_810* (-12.49, 95%-CI [-19.83, -5.15], p<0.001) resulted in a significant pleasantness decrease compared to the combination *_none*. Regarding the other combinations of structural resonances, only the addition of the other two structural resonances *_530* and *_710* to the resonance *_810* (*_530_710_810*) further decreased the pleasantness significantly (-8.82, 95%-CI [-16.16, -1.48], p=0.007), while the other combinations were not significantly different from each other.

Table 3. Same as Table 2, but now for the fl condition

Factor	Sum of Squares	Degrees of Freedom	Mean Square	F	p
Presence of 48th Order	8413.05	1	8413.05	36.14	<0.001
Level of 24th Order	80154.42	1	80154.42	344.54	<0.001
Combination of Structural Resonances	8178.04	7	8178.04	5.02	<0.001

The *cd* results (Table 4) differed from the other two conditions in a way that only the presence of the 48th order and the level of the 24th order had a significant effect on the pleasantness, while the occurrence of structural resonances did not significantly affect the pleasantness. On average, the presence of the 48th order reduced the perceived pleasantness from 47.3 to 37.9. If the level of the 24th order was 40 dB SPL, the average pleasantness was lower (33.7) than if the 24th order level was 25 dB SPL (51.5).

Table 4. Same as Table 2, but now the cd condition.

Factor	Sum of Squares	Degrees of Freedom	Mean Square	F	p
Presence of 48th Order	7150.44	1	7150.44	23.91	<0.001
Level of 24th Order	25573.05	1	25573.05	85.53	<0.001

To verify the results of the absolute magnitude estimation experiment, to show the relation to the variable MOTC and to quantify the perceptual ratio for both variables, a subset for each driving condition was chosen to conduct a pairwise comparison experiment for both variables. As a subset the sounds were chosen, which only contained the 24th order with a level of 25 dB SPL. This choice is justified in a way that in the majority of the cases only one order was audible and that an order level of 25 dB SPL is realistic within the vehicle interior. Additionally, the _ref stimulus was included in the subset for each driving condition.

Figure 7 shows the BTL scaled pleasantness results of the pairwise comparison experiment for the three driving conditions ru (left panel), fl (center panel) and cd (right panel) in comparison to the pleasantness results of the magnitude estimation experiment. The results show that for all driving conditions, the BTL scaled results are in good agreement with the magnitude estimation results, which is shown by correlation coefficients of 0.98 for the ru condition, 0.99 for the fl condition and 0.94 for the cd condition. For the ru condition, the BTL scaled pleasantness value of the _ref sound was 0.247. When a 24th order was added (ru_1_25_none), on average, the value decreased to 0.14, which is equivalent to a pleasantness reduction of 43 percent. When adding all structural resonances (ru_1_25_530_710_810), the BTL scaled pleasantness reduced to 0.024, which is equivalent to almost a factor of six compared to the sound ru_1_25_none without structural resonances. The implications for the other two conditions are similar, even though the ratios were different: For the fl condition, the pleasantness reduced from 0.407 to 0.183 when adding the 24th order and to 0.041 when adding all structural resonances. For the cd condition, adding the 24th order caused a pleasantness reduction from 0.613 to 0.122 and including both structural resonances resulted in a further pleasantness decrease to a value of 0.039. Note that the BTL scaled values are arbitrarily chosen by the algorithm and are therefore not comparable between the driving conditions or with values from other experiments.

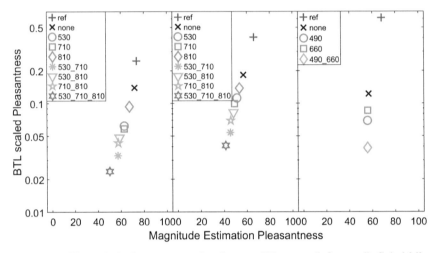

Figure 7. BTL scaled pleasantness for the conditions ru (left panel), fl (middle panel) and cd (right panel) plotted over the pleasantness from the magnitude estimation experiment. The symbols are the same as in Figures 4-6 with the exception that the grey crosses indicate the reference condition. Similar structure as in Doleschal et al., 2021 [59], but other data.

Figure 8 shows the BTL scaled MOTC values over the mean pleasantness values from the magnitude estimation experiment. The correlation coefficients of -0.94, -0.99 and -0.98 for the conditions ru, fl and cd between the BTL scaled MOTC and the pleasantness from the magnitude estimation experiment indicate a strong influence of the perceived MOTC on the pleasantness. The quantitative assessment of the BTL scaled MOTC values reveals that adding a 24th order to the reference noise increases the MOTC from 0.007 to 0.023, which is equivalent to a factor of 3.4. When adding all structural resonances, the MOTC further increases to 0.197. Similar results are also shown for the conditions fl and cd, even though the increase factors differ from the ru condition: For the fl condition, the MOTC increases from 0.006 to 0.034 when adding a 24th order and to 0.308 when adding all structural resonances. For the cd condition, adding the 24th order increases the BTL scaled MOTC from 0.015 to 0.145 and to 0.379 when adding both structural resonances.

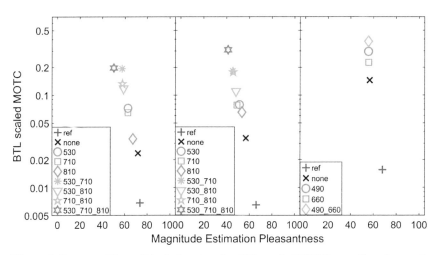

Figure 8. Same as Figure 7 but showing the BTL scaled MOTC over the pleasantness from the magnitude estimation experiment.

4.3 Generalized Synthetic Electric Vehicle Interior Sounds

To simplify the investigation of the influence parameters on pleasantness and MOTC, an experiment was conducted, where the structural resonances were generalized in the parameters modulation frequency and modulation index. The MOTC data have been previously published in Doleschal et al., 2021 [59]. Some of the aspects also previously have been discussed in Doleschal et al., 2020 [53], Doleschal et al., 2021 [59] and Doleschal and Verhey, 2022 [49].

4.3.1 Stimuli

To simplify the parameter variations of the experiments and to investigate the influences of different parameters more easily, in contrast to the previous experiment, the structural resonances were realized by applying a generalized amplitude modulation to the sinusoidal sweeps.

For that experiment, the background noise and the sweeps of the 24th and 48th order for the three driving conditions were realized in the same way as in the previous experiment with the difference that, besides 25 dB SPL and 40 dB SPL, the level of the unmodulated 24th order was also set to 15 dB SPL. This choice was motivated by the fact that for many recorded sounds the electric motor and gear orders are only audible when they pass

41

the frequency ranges of the structural resonances and are otherwise completely masked. According to the previous experiment, the level of the 48th order was always 6 dB lower than the one of the 24th order. The structural resonances were realized by multiplying the sinusoidal sweep according to equation (1) with an envelope $s_m(t, f_{mod})$ according to the equation

$$s(t) = s_c(t, f_b, f_e) \cdot (m \cdot s_m(t, f_{mod}) + 1) \tag{3}$$

where m denotes the modulation index, describing the intensity of the structural resonances. The envelope $s_m(t, f_{mod})$ was defined as

$$s_m(t, f_{mod}) = \left(sin\left(2\pi \cdot \frac{f_{mod}}{2} \cdot t \right) \right)^n, \tag{4}$$

with the modulation frequency f_{mod} and the exponent n. The exponent n was related to the modulation frequency f_{mod} to compensate for the temporal width of the structural resonances when increasing the modulation frequency. The relation was realized according to the following equation:

$$n = \frac{2.56}{f_{mod}^2} \tag{5}$$

This type of amplitude modulation was chosen since it mimics the amplitude modulations, which commonly occur in electric vehicle interior recordings. The values for the modulation frequency of the 24th order f_{mod} were either 0.2 or 0.4 Hz since these values are equivalent to one or two structural resonances during the signal duration of five seconds. For the 48th order, it was assumed that that order involves twice as many structural resonances as the 24th order since the frequency range of that order was twice as large. According to that, the modulation frequency of the 48th order was always double as high as that of the 24th order. The modulation index m was set to 2, 10 or 18. The values were chosen in a way that 2 and 10 represent typical values in the vehicle interior and 18 represents an extreme case. Due to the choice of positive and even values, the amplitude of the orders only increased during the structural resonances but did not increase outside them.

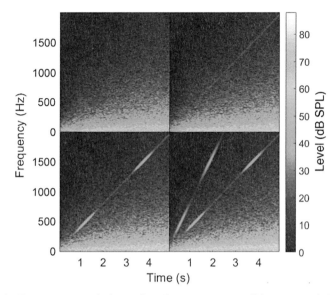

Figure 9. Parameter variations for the run-up condition: ru_ref (top left), ru_1_25_unmod, ru_1_25_0.4_10 (bottom left), ru_2_25_0.4_10 (bottom right). The color (from dark blue to yellow) indicates the sound pressure level over time and frequency. Adapted from Doleschal et al., 2021 [59].

The first three parts of the stimulus names were set according to the previous experiment (driving condition, number of orders and level of the 24th order). The fourth part of the sound name indicates the modulation frequency of the 24th order (_0.2 or _0.4) and the last part indicates the modulation index (_2, _10 or _18) of both orders. Additionally, six unmodulated stimuli (for each of the three levels with and without 48th order) were added, where the fourth and fifth part of the stimulus name was replaced by _unmod. To quantify the influence of tonal components, one noise-only stimulus (_ref) for each driving condition was added to the set. Combining all parameter variations with each other adds up to 43 sounds for each driving condition, in total 129 sounds for the whole experiment. Figure 9 shows spectrograms of an overview over the parameter variations. From top left (ru_ref) to top right (ru_1_25_unmod), a 24th order with a level of 25 dB SPL was added. The lower two subfigures show the addition of amplitude modulations (ru_1_25_0.4_10) and of a 48th order (ru_2_25_0.4_10).

43

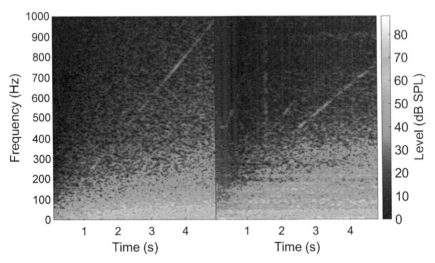

Figure 10. Left: Synthetic run-up sound containing a 24th electric motor order with a level of 25 dB SPL, a modulation frequency of 0.4 Hz and a modulation index of 2. Right: Spectrogram of a part-load run-up from 0-30 km/h recorded at the driver's left ear position. The color (from dark blue to yellow) indicates the sound pressure level over time and frequency. Adapted from Doleschal et al., 2021 [59].

Figure 10 shows a synthetic stimulus in comparison to an interior recording: The left panel displays the spectrogram of the sound *ru_1_25_0.4_2*, which contains a 24th electric motor order with a sound pressure level of 25 dB SPL with a modulation frequency of 0.4 Hz and a modulation index of 2. The right panel shows a spectrogram of a part-load run-up from 0-30 km/h of a compact car recorded at the driver's left ear position. The figure shows that the synthetic sound captures the key features of the recorded sound.

To assess whether the influence parameters driving condition, presence of the 48th order, level of 24th order, modulation frequency and modulation index influence the pleasantness and MOTC, both were measured in a magnitude estimation experiment. To verify the results and to quantify the impact of certain parameters on a ratio scale a subset was chosen for a pairwise comparison experiment. For the subset, the sounds with only the 24th order (_1) and a level of 25 dB SPL were chosen. For most of the recorded interior sounds, only one order is audible at the same time and a 24th-order level of 25 dB SPL is realistic in the vehicle interior. Additionally, the stimuli with an unmodulated 24th order and the noise-only sound _ref were added.

4.3.2 Results

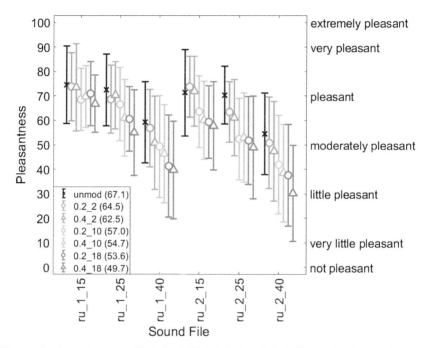

Figure 11. Mean values and interindividual standard deviations for pleasantness. Error bars indicate +/- one standard deviation from the mean. The labels on the abscissa indicate whether a 48th order was present (ru_2) or not (ru_1). The colors and symbols indicate whether structural resonances were present or not, their modulation frequency and their modulation index. Note that the reference sound _ref is not shown. The values in parentheses refer to the mean value of the pleasantness of all stimuli with the designated modulation parameters. Same structure as in Doleschal et al., 2021 [59], but pleasantness instead of MOTC data.

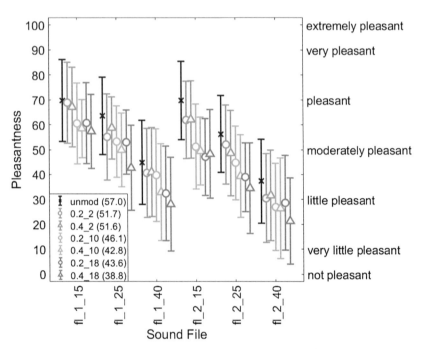

Figure 12. Same as Figure 11 but here for the fl condition. Same structure as in Doleschal et al., 2021 [59], but pleasantness instead of MOTC data.

In contrast to the other experiments, only 16 listeners participated in this experiment. Figures 11, 12 and 13 show the mean values and standard deviations for the variable pleasantness of the 16 listeners for the three driving conditions *ru, fl* and *cd*. For the *ru* condition, average pleasantness values range from 27.9 (*ru_2_40_0.4_18*) to 76.7 (*ru_ref*). For the *fl* condition the average pleasantness values range from 18.3 (*fl_2_40_0.4_18*) to 73.0 (*fl_ref*). For the *cd* condition average pleasantness values range from 22.1 (*cd_2_40_0.4_18*) to 73.5 (*cd_1_15_unmod*). Note that for the *cd* condition the sound *cd_1_15_unmod* was perceived as even more pleasant than the sound *cd_ref* (73.3).

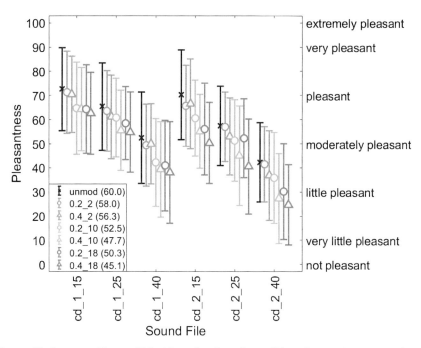

Figure 13. Same as Figure 11 but here for the cd condition. Same structure as in Doleschal et al., 2021 [59], but pleasantness instead of MOTC data.

Figures 14, 15 and 16 show the mean values and standard deviations for the variable MOTC of the 16 listeners for the three driving conditions *ru*, *fl* and *cd*. For the *ru* condition average MOTC values range from 3.9 (*ru_1_15_0.2_2*) to 79.3 (*ru_2_40_0.4_18*). For the *fl* condition the average MOTC values range from 4.9 (*fl_ref*) to 85.9 (*fl_2_40_0.4_18*). For the *cd* condition average MOTC values range from 3.5 (*cd_ref*) to 81.9 (*cd_2_40_0.4_18*). Note that for the *ru* condition the sound *ru_1_15_0.2_2* was perceived as less tonal than the sound *ru_ref* (4.4).

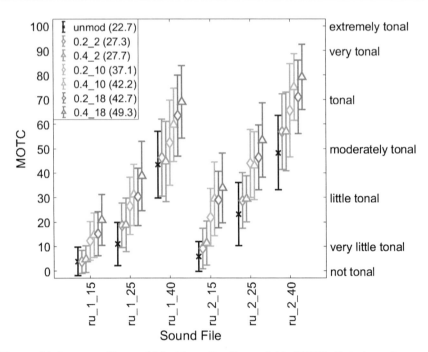

Figure 14. Same as Figure 11 but here for the variable MOTC. Data have been previously published in Doleschal et al., 2021 [59].

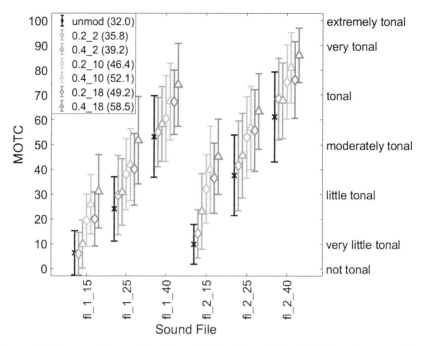

Figure 15. Same as Figure 14 but here for the fl condition. Data have been previously published in Doleschal et al., 2021 [59].

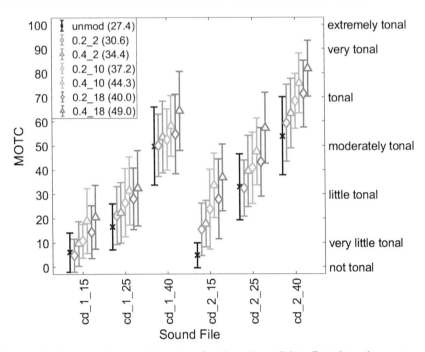

Figure 16. Same as Figure 14 but here for the cd condition. Data have been previously published in Doleschal et al., 2021 [59].

To evaluate which independent variables significantly influence the dependent variables pleasantness and MOTC, a multivariate analysis of variance (MANOVA) with a significance level of 5% was calculated. As independent variables, the following parameters were selected:

- Driving condition (*ru*, *fl* or *cd*)
- Presence of 48th order
- Level of 24th order
- Modulation frequency
- Modulation index

The significant results were summarized in Table 5. All of the previously mentioned independent variables had a significant effect on the combination of the variables pleasantness and MOTC. Furthermore, the interaction of the parameters presence of 48th order and level of 24th order and the interaction of modulation frequency and modulation index had a significant effect.

Table 5. Table of significant MANOVA results for the dependent variables pleasantness and MOTC. The independent variables were driving condition, presence of 48th order, level of 24th order, modulation frequency, modulation index and their interactions. Only significant interactions are shown. The table shows Wilk's Lambda (Λ), the F-value, the hypothesis and error degrees of freedom and the p-value for each independent variable or interaction.

Independent Variable	Λ	F	Hyp. df	Error df	p
Driving Condition	0.92	40.36	4	3868	<0.001
Presence of 48th Order	0.84	191.21	2	1934	<0.001
Level of 24th Order	0.35	677.28	2	1934	<0.001
Modulation Frequency	0.96	40.5	2	1934	<0.001
Modulation Index	0.81	109.22	4	3868	<0.001
Presence of 48th Order * Level of 24th Order	0.99	3.1	4	3868	0.015
Modulation Frequency * Modulation Index	0.99	3.63	4	3868	0.006

For the subsequent ANOVAs, a Bonferroni correction was applied. Univariate testing indicated that all independent variables had a significant influence on the dependent variables pleasantness (see Table 6) and MOTC (see Table 7). Furthermore, three interactions of independent variables had a significant influence on the MOTC.

Table 6. Table of significant ANOVA results of the dependent variable pleasantness for the independent variables driving condition, presence of 48th order, level of 24th order, modulation frequency and modulation index. All interactions were not significant and are therefore not shown.

Independent Variable	Type III Sum of Squares	df	Mean Square	F	p
Driving Condition	26310.14	2	13155.07	45.92	<0.001
Presence of 48th Order	25587.49	1	25587.49	89.33	<0.001
Level of 24th Order	185907.32	2	92953.66	324.52	<0.001
Modulation Frequency	4194.33	1	4194.33	14.64	<0.001
Modulation Index	33649.87	2	16824.93	58.74	<0.001

Table 7. Table of significant ANOVA results of the dependent variable MOTC for the independent variables driving condition, presence of 48th order, level of 24th order, modulation frequency and modulation index. Only significant interactions are shown.

Independent Variable	Type III Sum of Squares	df	Mean Square	F	p
Driving Condition	17908.63	2	8954.31	50.09	<0.001
Presence of 48th Order	61802.87	1	61802.87	345.7	<0.001
Level of 24th Order	613515.44	2	306757.72	1715.88	<0.001
Modulation Frequency	13541.53	1	13541.53	75.75	<0.001
Modulation Index	73367.91	2	36683.96	205.2	<0.001
Driving Condition * Modulation Index	2180.15	4	545.04	3.05	0.016
Presence of 48th Order * Level of 24th Order	2045.56	2	1022.78	5.72	0.003
Modulation Frequency * Modulation Index	2381.74	2	1190.87	6.66	0.001

The effect of the driving condition on the perceived pleasantness was highly significant. Tukey post-hoc tests indicated, that the pleasantness of the sounds of all driving conditions was significant from each other: *ru* vs. *fl* (10.96, 95%-CI [8.82, 13.10], p<0.001), *ru* vs. *cd* (5.56, 95%-CI [3.42, 7.70], p<0.001), *fl* vs. *cd* (-5.40, 95%-CI [-7.54, -3.26], p<0.001). On average, the pleasantness of the sounds of the *ru* condition was 58.8, followed by the *cd* condition (53.3) and the *fl* condition (47.9). In addition, the presence of the 48th order had a significant effect on the pleasantness. Adding a 48th order decreased, on average, the pleasantness from 56.8 to 49.0. An increase of the 24th order level significantly reduced the average pleasantness from 63.9 (15 dB SPL) to 55.3 (25 dB SPL) and 39.5 (40 dB SPL): _15 vs _25 (8.58, 95%-CI [6.16, 11.00], p<0.001), _15 vs. _40 (24.41, 95%-CI [21.98, 26.83], p<0.001), _25 vs. _40 (15.82, 95%-CI [13.40, 18.25], p<0.001). The modulation frequency, which indicates the number of structural resonances during the signal duration of five seconds, also had a significant influence. Increasing the modulation frequency from 0.2 to 0.4 Hz results in an average pleasantness decrease from 53.0 to 49.9.

This also holds true for the modulation index. Tukey post-hoc tests revealed significant differences between all modulation indexes: 2 vs. 10 (7.26, 95%-CI [4.69, 9.82], p<0.001), 2 vs. 18 (10.57, 95%-CI [8.00, 13.13], p<0.001), 10 vs. 18 (3.31, 95%-CI [0.75, 5.87], p=0.005). An increase in the modulation index from 2 to 10 reduced, on average, the pleasantness from 57.4 to 50.1. A further increase to 18 resulted in a further pleasantness reduction to 46.8.

All the investigated independent variables also had a significant effect on the MOTC. For the driving condition, a Tukey post-hoc test revealed that the perceived MOTC between the three conditions was significantly different: ru vs. fl (-8.98, 95%-CI [-10.67, -7.29], p<0.001), ru vs. cd (-1.95, 95%-CI [-3.64, -0.26], p<0.001), fl vs. cd (7.03, 95%-CI [5.33, 8.72], p<0.001). On average, the MOTC of the fl sounds was 43.8, followed by the cd condition (36.8) and the ru condition (34.9). The effect of adding a 48th order also had a significant effect on the MOTC. Sounds, which contained both orders, were perceived as more tonal (45.4) than sounds with only the 24th order (33.2). In addition, the level of the orders played an important role in the MOTC perception. The comparisons between the levels _15 and _25 (-17.72, 95%-CI [-19.66, -15.78], p<0.001), _15 and _40 (-44.26, 95%-CI [-46.20, -42.32], p<0.001) and _25 and _40 (-26.54, 95%-CI [-28.48, -24.60], p<0.001) reveal that an increase of the level leads to an increase of the MOTC. If the level of the 24th order was increased from 15 dB SPL to 25 dB SPL, the mean MOTC increased from 18.6 to 36.4. If the level was further increased to 40 dB SPL, the mean MOTC further increased to 62.9. The effect of the parameters modulation frequency and modulation index had a further effect on the perceived MOTC. Doubling the modulation frequency from 0.2 to 0.4 Hz led to an average increase in the MOTC from 38.5 to 44.1. For the modulation index, the Tukey post-hoc test showed significant increases of the MOTC between all pairwise comparisons: 2 vs. 10 (-10.72, 95%-CI [-12.82, -8.63], p<0.001), 2 vs. 18 (-15.59, 95%-CI [-17.69, -13.50], p<0.001), 10 vs. 18 (-4.87, 95%-CI [-6.97, -2.78], p<0.001).

Comparing the BTL scaled pleasantness data with the magnitude estimation pleasantness data results in correlation coefficients of 0.96, 0.96 and 0.98 for the conditions ru, fl and cd. The high correlation coefficients reveal that the data from both experiments are in good agreement. The results further show that the subset captures a high dynamic range of the variable pleasantness. For the *ru* condition, the most pleasant sound *ru_ref* is 9.8 times more pleasant than the sound *ru_1_25_0.4_18*. Similar ratios occur for the other conditions: For the *fl* condition, the most pleasant sound *fl_ref* is nine times more pleasant than the sound *fl_1_25_0.4_18* and for the cd condition the sound *cd_ref* is 7.2 times more pleasant than the sound *cd_1_25_0.4_18*.

For the MOTC, the BTL scaled data are also in good agreement with the magnitude estimation data, which is shown by correlation coefficients of 0.98, 0.98 and 0.99 for the conditions *ru*, *fl* and *cd*. The most tonal sounds were the least pleasant sounds for all conditions and vice versa, even though the dynamic ranges were considerably higher than for the pleasantness, shown by the values of 34.4 (*ru*), 41.2 (*fl*) and 51.9 (*cd*).

4.4 Acoustic Optimization with Synthetic Sounds

The following section describes the acoustic optimization experiment with synthetic sounds to gain first insights into the influence of different types of subharmonics and the sound pressure level of the background noise on the pleasantness and MOTC. The experimental data were presented by Doleschal and Verhey, 2022 [49]. The discussion here is also taken in part from this paper.

4.4.1 Stimuli

The stimuli were created in a similar way as in the previous experiment: The sounds of the three conditions *ru*, *fl*, and *cd* (first part of the stimulus name) contained either only the 24th (_1) or both the 24th and 48th (_2) electric motor orders, indicated by the second part of the stimulus name. As before, the orders were calculated according to the equations (1), (3), (4) and (5), with the difference, that the modulation frequency was fixed to a value of 0.4 Hz and the modulation index to a value of 2. These values were chosen, as they are realistic within the vehicle interior. In addition, the level of the 24th order was fixed to a value of 40 dB SPL, which results in a level of 34 dB SPL for the 48th order if the stimulus contained both orders. A 24th-order level of 40 dB SPL is higher than typical in the vehicle interior. This value was chosen to make the order audible during a large temporal portion of the stimulus and to make the effects of potential pleasantness improvements due to acoustic optimization measures visible.

In contrast to the previous experiments, the influence of the background noise was additionally evaluated. Therefore, the maximum noise level was altered. The higher maximum noise level was the same as in the previous experiment, while the lower one was 10 dB SPL below the higher level. For the sounds of the *ru* and *fl* conditions, the noise level reached its maximum at the end of the stimulus. For the stimuli of the *cd* condition, the noise level was highest at the beginning. For the sounds of the *ru* condition, the background noise level increased from 27 to either 67 (_67) or 77 dB SPL (_77), which is denoted by the third part of the stimulus name. For the *fl* condition, the background noise level increased from 27 to 73 dB SPL (_73) or 83 dB SPL (_83). For the stimuli of the *cd* condition, the background noise level reduced from 82 (_82) to 72 dB SPL or from 72 (_72) to 62 dB SPL.

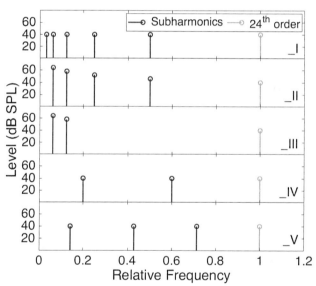

Figure 17. Spectra of tonal content containing the 24th electric motor order (light brown) and the sets of subharmonics according to Gwak et al., 2014 [47] (blue). The labels _I to _V in the lower right corner of each panel refer to the last part of the sound name and are used to differentiate between the above-shown spectra. The abscissa shows the frequency of each component relative to the frequency of the 24th order. The ordinate shows the level of each component. Figure previously published as Figure 1 in Doleschal and Verhey, 2022 [49].

The last part of the stimulus name denotes the spectrum of tonal content, i.e. the subharmonics. The spectra, which are shown in Figure 17, were adapted from Gwak et. al. [47] and are labelled by an underscore, followed by a Roman number. The frequencies of the subharmonics were set relative to the frequency of the 24th order. Three different types of subharmonics were used: The octave-spaced subharmonics of the spectra _I, _II and _III were motivated in a way that the usage of an octave spacing leads to the highest consonance of the resulting complex tones. The odd-numbered subharmonics of the spectra _IV and _V imitate the spectrum of a closed-pipe musical instrument [47] and contain the consonant intervals perfect fifth and major sixth.

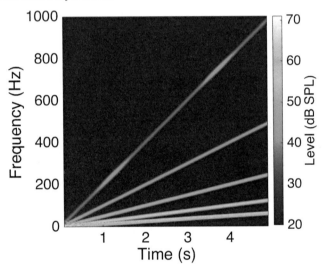

Figure 18. Spectrogram of the sound ru_1_67_II without background noise, showing the 24th order and four subharmonics, where the frequency was halved towards lower frequency. The level increased by 6 dB towards lower-frequency subharmonics. Figure adapted from Doleschal and Verhey, 2022 [49], but another spectrum of tonal content.

The spectrum of tonal content _I (top panel of Figure 17) contained five subharmonics, where the frequency was halved towards lower-frequency subharmonics. The level of all subharmonics was the same as the level of the 24th order, i.e. 40 dB SPL. The stimuli with the spectra of tonal content _II (shown in Figure 18) contained four subharmonics, where the frequency was halved towards lower frequency. The level of each subharmonic increased by 6 dB towards lower-frequency subharmonics, so the level of the lowest-frequency subharmonic was 64 dB SPL. The motivation to include this spectrum was to compensate for the increase in masking due to the low-pass characteristic of the background noise. The spectrum of tonal content _III contained the two lowest-frequency subharmonics of spectrum _II with levels of 58 and 64 dB SPL. This spectrum has the advantage to shift the fundamental frequency downwards while the loudness increase due to the addition of subharmonics is lower than when using the spectra _I or _II.

The spectra of tonal content _IV and _V contained odd-numbered subharmonics with component ratios of 1:3:5 and 1:3:5:7, where all subharmonics had the same level as the 24th electric motor order, i.e. 40 dB SPL. The motivation to include these spectra was the lower number of subharmonics compared to the octave-spaced spectra so the loudness increase due to the addition of subharmonics was lower than for the octave-spaced spectra.

As a reference, a spectrum of tonal content was included, which consisted of only electric motor orders and did not contain any subharmonics. This spectrum is referred to as _0. Combining all parameters with each other (presence/absence of 48th order, maximum noise level and spectrum of tonal content) results in 24 sounds per driving condition, 72 sounds in total.

4.4.2 Results

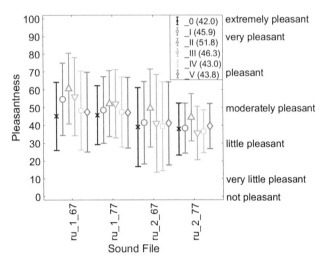

Figure 19. Mean values and standard deviations of the acoustic optimization experiment of the 20 listeners who participated in the experiment. Error bars indicate +/- one standard deviation of the mean. The data are grouped into subsets of six data points with the same combination of the parameters presence/absence of 48th order and maximum background noise level. The values in parentheses indicate the mean value of all values with the respective spectrum of tonal content of the ru condition. Data have been previously published in Doleschal and Verhey, 2022 [49].

The following part describes the results of the 20 listeners, who participated in the magnitude estimation experiment. Figure 19 shows the mean values and interindividual standard deviations of the *ru* condition. The error bars indicate +/- one standard deviation from the mean. The data are grouped by the parameters presence of 48th order and maximum background noise level to enable an easy comparison of the different spectra of tonal content. For the *ru* condition, average pleasantness values range from 35.7 to 60.7, which is the smallest dynamic range among the three conditions.

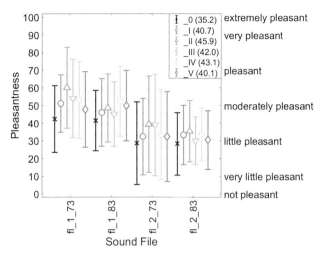

Figure 20. Same as Figure 19 but here for the fl condition. Data have been previously published in Doleschal and Verhey, 2022 [49].

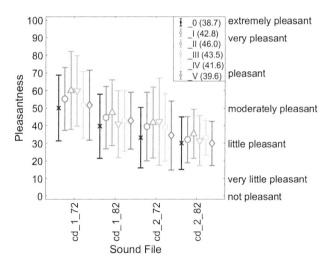

Figure 21. Same as Figure 19 but here for the cd condition. Data have been previously published in Doleschal and Verhey, 2022 [49].

Figure 20 shows the pleasantness ratings for the sounds of the *fl* condition. With a range from 28.3 to 60.1, the ratings of the *fl* condition span a wider range than those of the *ru* condition.

Figure 21 shows the average pleasantness ratings of the *cd* condition. The mean pleasantness of this condition ranges from 29.8 to 59.9. To verify the results of the magnitude estimation experiment a subset was chosen, which only contained the sounds, where only the 24th order was present. The correlation coefficients of 0.67 (*ru*), 0.88 (*fl*) and 0.93 (*cd*) show, that the BTL scaled results are generally in agreement with the magnitude estimation results. The lower correlation coefficient of the *ru* condition results from the lower dynamic range of the *ru* condition compared to the other two conditions. The dynamic range of the subset, which contained only the 24th electric motor order, was considerably lower than that of the whole dataset, ranging from 45.2 to 60.7. The relatively small dynamic range is also represented by the BTL scaled values: from 0.055 to 0.127 for the *ru* condition (factor 2.3), from 0.064 to 0.128 for the *fl* condition (factor 2.0) and from 0.042 to 0.156 (factor 3.7) for the *cd* condition.

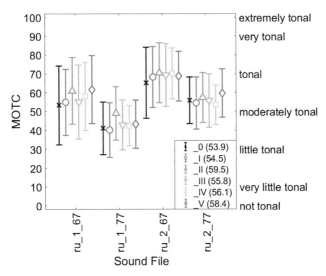

Figure 22. Same as Figure 19 but here for the perceived MOTC of the ru condition. Data have been previously published in Doleschal and Verhey, 2022 [49].

Figure 22 shows the average MOTC ratings for the *ru* condition. The ratings of this condition range from 40.2 to 70.6. The MOTC results of the *fl* condition, which are shown in Figure 23, were similar to those of the *ru* condition. With a range from 48.1 to 79.6, the results of the full-load run-up condition span a slightly larger range than those of the *ru* condition.

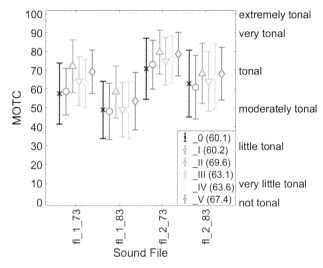

Figure 23. Same as Figure 22 but here for the fl condition. Data have been previously published in Doleschal and Verhey, 2022 [49].

The MOTC results of the *cd* condition, shown in Figure 24, span the largest range among the three conditions. Mean MOTC data range for that condition range from 39.9 to 79.5.

For the MOTC, the same subset as for the pleasantness was selected to conduct a pairwise comparison experiment. The correlation coefficients between the magnitude estimation experiment and the BTL scaled data are 0.90 (*ru*), 0.86 (*fl*) and 0.90 (*cd*) and show that the data of both experiments are in good agreement. The BTL scaled MOTC data range from 0.052 to 0.136 (factor 2.6), from 0.041 to 0.193 (factor 4.7) for the *fl* condition and from 0.046 to 0.249 (factor 5.4). They span a larger range than the pleasantness data.

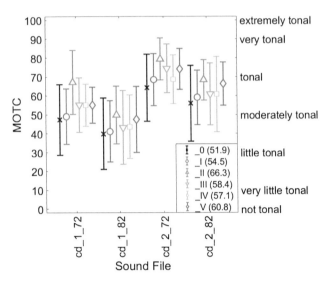

Figure 24. Same as Figure 22 but here for the cd condition. Data have been pre-viously published in Doleschal and Verhey, 2022 [49].

Because the dependent variables pleasantness and MOTC are partially correlated ($r = -0.34$), a multifactorial MANOVA with a significance level of 5% was carried out. Therefore, the following independent variables were included in the MANOVA:

- Driving condition (*ru*, *fl*, *cd*)
- Presence of 48th order
- Maximum background noise level (categorized as low or high)
- Spectrum of tonal content (_0 to _V)

Note that even though the level over time differs for the three conditions *ru*, *fl* and *cd*, the maximum noise levels always had a difference of 10 dB SPL. Therefore, the variable maximum background noise level was categorized as low or high for the three conditions. The significant results of the MANOVA are shown in Table 8. All investigated variables and two interactions revealed a significant influence on the combination of the variables.

Table 8. Multivariate analysis of variance for the dependent variables pleasant-ness and MOTC for the independent variables driving condition, presence of 48th order, maximum background noise level, spectrum of tonal content and their interactions. All independent variables show a significant effect on the combi-nation of the dependent variables pleasantness and MOTC. Furthermore, two interactions revealed a significant influence. Non-significant interactions were omitted.

Independent Variable	Λ	F	Hyp. df	Error df	p
Driving Condition	0.95	19.17	4	2734	<0.001
Presence of 48th Order	0.75	226.94	2	1367	<0.001
Maximum Background Noise Level	0.84	126.26	2	1367	<0.001
Spectrum of Tonal Con-tent	0.93	10.47	10	2734	<0.001
Driving Condition * Presence of 48th Order	0.99	3.58	4	2734	0.006
Driving Condition * Maximum Background Noise Level	0.99	2.38	4	2734	0.050

To indicate whether the independent variables influence each variable, univariate ANOVAs were conducted. To compensate for the type I error accumulation, the Bonferroni correction was applied. The univariate ANO-VAs indicated that all independent variables had a significant influence on each of the dependent variables pleasantness (Table 9) and MOTC (Table 10).

Table 9. Table of significant ANOVA results for the pleasantness. As independent variables, the parameters driving condition, presence of 48th order, maximum background noise level and spectrum of tonal content were considered. Interactions were not significant and are therefore not shown.

Independent Variable	Type III Sum of Squares	df	Mean Square	F	p
Driving Condition	5000.53	2	2500.26	6.61	0.001
Presence of 48th Order	64400.89	1	64400.89	170.2	<0.001
Maximum Background Noise Level	9664.95	1	9664.95	25.54	<0.001
Spectrum of Tonal Content	11723.7	5	2344.74	6.2	<0.001

For the parameters driving condition and spectrum of tonal content, Tukey post hoc tests were calculated to check which mean values significantly differ from each other. For the driving condition, significant pleasantness differences were found between the driving conditions *ru* and *fl* (4.29, 95%-CI [1.14, 7.23], p=0.002) and between the conditions *ru* and *cd* (3.5, 95%-CI [0.56, 6.45], p=0.015). The sounds of the *ru* condition had an average pleasantness of 45.4, while the *fl* and *cd* sounds, on average, had pleasantness ratings of 41.2 and 41.9. The ANOVA further revealed that the differences between the sounds with and without a 48th order were significant. Sounds without a 48th order, on average, had a pleasantness rating of 49.5, which dropped to 36.2, if a 48th order was present. The maximum background noise level also significantly influenced the pleasantness. An increase from the lower to the higher value, on average, led to a decrease in pleasantness from 45.4 to 40.3. When regarding the spectrum of tonal content, based on a significance level of 5%, the spectrum _0 without subharmonics significantly differed from the spectra _II (-9.44, 95%-CI [-14.51, -4.37], p<0.001) and _III (-5.47, 95%-CI [-10.54, -0.4], p=0.026). Spectrum _II also differed from spectra _IV (5.34, 95%-CI [0.27, 10.4], p=0.032) and _V (6.71, 95%-CI [1.64, 11.78], p=0.002).

Table 10. Same as Table 9 but now for the dependent variable MOTC.

Independent Variable	Type III Sum of Squares	df	Mean Square	F	p
Driving Condition	15260.3	2	7630.15	33.51	<0.001
Presence of 48th Order	70802.83	1	70802.83	310.91	<0.001
Maximum Background Noise Level	49398.29	1	49398.29	216.92	<0.001
Spectrum of Tonal Content	16024.15	5	3204.83	14.07	<0.001
Driving Condition * Presence of 48th Order	1884.05	2	942.02	4.14	0.016

Similarly, to the dependent variable pleasantness, Tukey post hoc tests were carried out for the dependent variable MOTC for the variables driving condition and spectrum of tonal content, where more than two mean values were compared. For the driving condition, the MOTC was significantly different when comparing the conditions *ru* and *fl* (-7.63, 95%-CI [-9.91, -5.34], p<0.001) and the conditions *fl* and *cd* (5.83, 95%-CI [3.55, 8.12], p<0.001). The mean MOTC of the *fl* condition was significantly higher (64.0) than of the conditions *ru* (56.4) and *cd* (58.2). For the spectrum of tonal content, the following pairwise comparisons led to a significant increase of the MOTC: _*II* vs. _*0* (9.85, 95%-CI [4.37, 14.51], p<0.001), _*V* vs. _*0* (6.90, 95%-CI [2.97, 10.83], p<0.001), _*III* vs. _*II* (6.03, 95%-CI [2.10, 9.96], p<0.001) and _*II* vs. _*IV* (6.19, 95%-CI [2.26, 10.12], p<0.001). The results show that the average MOTC of the sounds with the most pleasant spectrum of tonal content was also rated as the most tonal.

4.5 Discussion

The results of the first experiment for electric vehicle interior sounds revealed that for all driving conditions, the independent variables presence of 48th order and level of 24th order significantly influenced the pleasantness. In contrast, the combination of structural resonances did not have a significant effect on the pleasantness for the *cd* condition. It is likely, that this difference did not result from the decreasing frequency of the sinusoidal sweep, but from the realization of the structural resonances. Observations from interior recordings showed that structural resonances are generally less audible within coast-down conditions, even though this does not apply to all recordings. For the conditions *ru* and *fl*, the exact center frequencies of the resonances do not seem as relevant as the number of resonances. Furthermore, the different realizations of the structural resonances of the three driving conditions do not allow comparing the pleasantness of the driving conditions with each other, even though the results indicate that the sounds of the *ru* were perceived as more pleasant than those of the *fl* and the *cd* conditions.

Regarding the experiment with generalized synthetic sounds, the results showed that the masked thresholds of the orders likely had an influence on both the variables pleasantness and MOTC. The 24th order with a level of 15 dB SPL was unlikely to be audible outside the structural resonances for a vast majority of the listeners, since, for all driving conditions, the previous experiment on masked thresholds revealed, that for all listeners, the average individual masked thresholds of the 24th order for the three conditions were above 15 dB SPL. Similar findings result from the thresholds of the 48th order: For most of the listeners, the level of the 48th order was below the masked threshold so that outside the structural resonances the order could not have had any effect, neither on the pleasantness nor on the MOTC. The results, therefore, reveal that the results for both pleasantness and MOTC for the sounds with the unmodulated orders with a level of 15 dB SPL are close to those of the _ref sounds. Note that the experimental masked threshold is a conservative estimation since the final threshold estimate only indicates that the order was audible at one moment during the signal duration. Considering the sounds with a 24th order level of 15 dB SPL, containing structural resonances with a modulation index of 2, which is equivalent to a maximum level increase of approximately 9.5 dB SPL, the experimental pleasantness values decreased and the MOTC values increased since the order level was above its masked threshold during the structural resonances. A further increase of the modulation index to 10 or 18, therefore, resulted in a further pleasantness decrease and MOTC increase, since the duration and the prominence of the audibility of the order increased.

The results could be generalized to the sounds with higher 24th-order levels: Even though for sounds with 24th-order levels of 25 and 40 dB SPL (and 48th-order levels of 19 and 34 dB SPL), the prominence was considerably higher compared to the sounds with a 24th-order level of 15 dB SPL, resulting in lower pleasantness and higher MOTC, increasing the modulation frequency and the modulation index still resulted in lower pleasantness and higher MOTC due to the prominence increase of the order.

The significant influence of the independent variables on the dependent variables pleasantness and MOTC could be explained in the light of several previously conducted studies: For both the investigated dependent variables pleasantness and MOTC, the values significantly differed between the driving conditions *ru*, *fl* and *cd*. One explanation for the different perceptions could be the different maximum frequencies of both simulated electric motor orders. For the *ru* condition, the maximum frequency of the 24th order was 1000 Hz, but 1600 Hz for the *cd* condition and 2000 Hz for the *fl* condition. Therefore, the slightly higher MOTC of the *cd* condition and the considerably higher MOTC of the *fl* condition could be explained by the different maximum frequencies. Vormann et al., 2000 [75] conducted a study for pure tones embedded into a uniform exciting noise, where the tone frequency was varied within a range between 200 and 3700 Hz. In that study, the perceived MOTC increased as the frequency of the tone increased. Therefore, the difference in the perceived MOTC between the three driving conditions could be explained in the light of a difference in maximum frequencies.

The study of Hellman, 1985 [76] confirmed the influence of the 48th order on perceived pleasantness: In that study, two-tone complexes were embedded into a low-pass background noise. The study showed that for sounds with an overall sound pressure level between 75 and 80 dB SPL, which is similar to the overall sound pressure level of the sounds presented in the current study, the perceived annoyance increased with an increasing spectral distance of the two tones if the spectral distance was 250 Hz or higher. This could be explained in a way that only for spectral distances of 250 Hz or more, but not for spectral distances of 25 Hz and 100 Hz, the components of the two-tone complexes were perceived as separate tones. Instead, the two-tone complexes with spectral distances of 25 Hz and 100 Hz were likely perceived as one amplitude-modulated tone, eliciting the sensations of fluctuation strength and roughness [8]. The results of the present experiment are in line with that study in a way that

67

the addition of a 48th order significantly reduced the pleasantness of the presented stimuli.

Landström et al., 1995 [77] conducted a study on the effect of tonal components on annoyance in working environments. The spectrum of the majority of the working environments mainly consisted of low-frequency components below 200 Hz and mid-frequency components between 200-2000 Hz, similarly to the investigated sounds of this experiment. They observed an increase in annoyance when more than one tonal component was audible compared to working environments where only a single tone was audible.

Vormann et al., 2000 [78] conducted a similar study investigating the influence of the number of higher harmonics on the perceived MOTC. They concluded that the addition of one higher harmonic strongly increased the MOTC compared to the condition without higher harmonics, while the effect of the addition of further higher harmonics was considerably smaller. Their outcomes confirm the present findings in a way that the addition of a 48th electric motor order strongly affects the perceived MOTC.

The influence on the level on perceived pleasantness was investigated in the study of Hellman, 1984 [79]. In that study, single tones of different frequencies were embedded into either a low-pass or a high-pass noise. The study revealed that when high-frequency tones were embedded into a low-pass noise, the annoyance strongly grew if the tone-to-noise ratio was increased. The results of the present study are consistent with these findings since an increase of the 24th order level strongly reduced the pleasantness. The decrease in pleasantness due to a level increase of the 24th order is also in line with the above-mentioned study of Landström et al., 1995 [77]. They showed that adding audible tonal components increases the annoyance and that a higher tonal component level leads to a higher annoyance. The data of the experiment on generalized synthetic electric vehicle sounds revealed that for components, whose level is above the masked threshold, the pleasantness considerably decreased and a further level increase led to a further pleasantness reduction.

Hansen and Weber, 2008 [80] investigated the influence of the tonal component level on the perceived MOTC of vehicle interior sounds. They conducted an experiment on the MOTC of vehicle interior sounds recorded in a coast-down condition, which contained an audible howling component.

They concluded that the tone-to-noise ratio of the howling component was directly related to the howling perception: A reduction of that tone-to-noise ratio caused a reduction of the perceived howling by a constant factor. Hansen et al., 2005 [81] pointed out that the adjective pair "tonal" and "not tonal" is suitable to characterize vehicle interior sounds containing "whistle" and "whining" sound components. Therefore, the results of the above-mentioned experiment agree with that study in a way that an increase of the 24th order level increases the perceived MOTC in case the level is above the masked threshold.

The MOTC further increased in case the parameters modulation frequency and modulation index were increased since in those cases the duration and intensity of the perceptibility of the tonal components were increased. The latter relation is further supported by the study of Hansen and Weber, 2008 [80] in a way that with an increasing duration the perception of the tonal content increases.

When regarding the parameters modulation frequency and modulation index, the increase of both variables also reduced the perceived pleasantness, which could be explained in the light of the study of Hellman, 1984 [79]. The increase of the variables modulation frequency and modulation index resulted in an increase in tone-to-noise ratio during the structural resonances and therefore increased the duration and intensity of the audibility of the tonal components.

Altogether, the previously discussed studies reveal that the significant influences of all the investigated independent variables driving condition, presence of 48th order, 24th order level, modulation frequency and modulation index on the pleasantness and MOTC are in line with previously conducted psychoacoustic experiments. Therefore, the results of the generalized experiment with synthetic sounds could be used as a basis for the following experiment on the acoustic optimization with synthetic sounds, the selection of the stimuli for the experiments on recorded electric and hybrid vehicle interior sounds and the data augmentation experiments for specific sound components.

The following part discusses the implications of the acoustic optimization experiment with synthetic sounds. As mentioned in section 4.3 for the experiment on generalized synthetic electric vehicle interior sounds, the studies of Landström et al., 1995 [77] and Hellman, 1985 [76] showed that sounds containing two tonal components are generally perceived as less pleasant, which seems to be at odds with the result that the addition of subharmonics has the potential to increase the pleasantness. However, in the study of Landström et al., 1995 [77], the frequency ratio of the tonal components was not further investigated so their findings, that the existence of multiple tonal components generally results in a pleasantness decrease does not contradict the present results. In the study of Hellman, 1985 [76], the frequency of the lower-frequency component of each two-tone complex was set to 250 Hz and only the higher-frequency component was varied. The separate audibility of the higher-frequency component increased the perceived annoyance, but the addition of a lower-frequency component was not investigated throughout the study.

Two studies explicitly investigated the influence of subharmonics on the pleasantness of vehicle interior sounds. One of them is the study of Gwak et al., 2014 [47], which motivated the spectra of tonal content used in both the optimization experiments for synthetic and recorded sounds. Their study distinguished between the variables pleasantness and preference rating, even though both variables were highly correlated so that the implications for the variable preference ratings are also partially valid for the pleasantness. In contrast to their study, the present results, on average, did not show any pleasantness reduction for spectra of tonal content with subharmonics compared to the one without, while in their study the preference rating of the sounds with the spectra _I, _II and _V was lower than for the sounds without subharmonics. In addition, the preferred spectra of both studies were different. While in the study of Gwak et al., 2014 [47] the listeners preferred the spectra _III and _IV, in the present study spectrum _II was preferred. The differences could result from the different frequencies of the sinusoidal sweep. While in the previous study, the underlying sweep ranged from 2800 Hz to 5200 Hz, the sweeps of the present study had considerably lower frequencies.

Another study about the application of subharmonics to improve the vehicle interior sound pleasantness was conducted by Sun et al., 2018 [48]. They extracted both a time-invariant and a time-variant whistle component from the same vehicle interior sound with a duration of five seconds. While the time-invariant whistle component had a constant center frequency of almost exactly 5000 Hz and had several side bands (likely resulting from the inverter), the time-variant whistle component only consisted of one to-nal component, which increased approximately linearly from 3150 Hz to 4150 Hz. Afterwards, they reconstructed the extracted components and added spectra of subharmonics with either octave-, fifth-, fourth-, odd-, or even-spaced subharmonics to that component. In contrast to the study of Gwak et al., 2014 [47] (and the present study), where the level of the sub-harmonics increased or was kept constant towards lower frequency, Sun et al., 2018 [48] reduced the level towards lower frequency. For the time-invariant and the time-variant whistle sounds, all the sounds, which con-tained subharmonics, were, on average, perceived as less annoying than those without. Comparing the different spectra containing subharmonics, for both the time-invariant and the time-variant whistle sounds, the sounds with the octave-spaced and the even-spaced subharmonics were per-ceived as least annoying. These results of the present study agree with the findings in a way that the sounds with the spectrum of tonal content _II with octave-spaced subharmonics were, on average, perceived as most pleasant, while spectra with even-spaced subharmonics have not been evaluated in the present experiment. From the comparison between the data of the previous and the present study on subharmonics, one could argue, that the level of the subharmonics is not relevant for the pleasant-ness perception. However, their study used the extracted components without background noise. Due to that reason, they did not have to con-sider the masking of the added subharmonics. Added subharmonics might be inaudible if their level is chosen too low.

In the acoustic optimization experiment with synthetic sounds, also the im-pact of the background noise was investigated. To investigate its impact, two different maximum background noise levels were evaluated. In gen-eral, the sounds with the lower maximum background noise level were perceived as more pleasant than those with the higher one, which could result from two different reasons: Firstly, a higher background noise level generally results in a higher loudness of the total sound. As an example, the sound $ru\_1\_67\_0$, a sound of the run-up condition, which only con-tained the 24th order and the background noise with the lower maximum level of 67 dB SPL and did not contain any subharmonics, has an average loudness of 1.78 sone according to the dynamic loudness model (DLM) of Chalupper and Fastl, 2002 [82]. The increase of the maximum background noise level by 10 dB (sound $ru\_1\_77\_0$) resulted in a considerably higher loudness of 3.36 sone. Due to the generally strong influence of loudness on the perceived pleasantness of vehicle interior sounds as shown in section 7.4 and in several other studies (e. g. Steinbach and Altinsoy, 2019 [41], Ma et al., 2017 [39]), the increase of loudness likely overweighed the reduction of the MOTC.

Another aspect of a higher background noise level was the masking of subharmonics. Due to the low-pass shape of the noise spectrum, even though an increase of the background noise also resulted in a masking of the 24th and 48th electric motor orders, it primarily affected the lower-frequency subharmonics. Due to the reason, that the perceived pitch of the tonal component is determined by the lowest-frequency tonal component, a higher noise level will cause a higher pitch perception. This implication further explains, why the sounds with the spectrum of tonal content _II were perceived over those with the spectrum _I, even though the mean loudness according to the DLM was higher, e.g. ru_1_67_2: 2.86 sone vs. ru_1_67_1: 1.99 sone. One could argue that an alternative could be to use a flat-spectrum or a high-pass noise. Even though the masking of the lower-frequency subharmonics would reduce, the increase of the loudness (and probably the sharpness) would take a largely negative effect on the pleasantness [39, 41]. Furthermore, Bodden and Belschner [55] argued that artificial sound components have to sound, as they would originate from the vehicle itself. The background noise of all of the investigated vehicles had a low-pass characteristic so the addition of a differently shaped noise would expose that the noise component does not originate from the car.

Altogether, the results show that the addition of subharmonics is a suitable approach to improve the pleasantness of vehicle interior sounds containing tonal components, even though they are perceived as louder as and more tonal than those without. In contrast, the experiment revealed, that an increase in the background noise level does not seem an adequate approach to increase the pleasantness due to the large loudness increase and the increased masking of lower-frequency subharmonics.

5 Experiments with Recorded Vehicle Interior Sounds

This chapter describes, in contrast to chapter 4, experiments on the perception of real recordings and their alterations.

5.1 Pure-Electric Driving Condition

5.1.1 Stimuli

To classify the recordings and to conclude which factors play a role in the perception of pleasantness, several non-acoustic parameters, e.g. the rotational speed of the electric motor [28] and the vehicle speed, were additionally acquired. The stimuli were extracted by cutting out binaural snippets with a duration of five seconds, which include the left and right channels from the driver's seat. Generally, the same driving conditions run-up (*ru*), full-load run-up (*fl*) and coast-down (*cd*) as in the experiments with synthetic sounds were distinguished. Furthermore, the signal snippets were cut in such a way, that the vehicle speed at the beginning of the *ru* and *fl* snippets was either 0, 30 or 50 km/h. For the *cd* condition, the vehicle speed at the beginning was either 70, 50 or 30 km/h.

For the experiment, 145 snippets were cut, which are recorded from ten different vehicles (recordings presented in [2]). Note that also hybrid electric vehicles with switched-off combustion engines were used to generate electric vehicle stimuli, as they can be driven in pure-electric mode. From the 145 snippets, 65 were from the *ru* condition, 20 from the *fl* condition and 60 from the *cd* condition.

The 65 sounds of the *ru* condition were further separated into the following speed ranges:

- 29 run-up sounds from 0 km/h
- 24 run-up sounds from 30 km/h
- 12 run-up sounds from 50 km/h

The choice was motivated by the reason that the recordings showed that the audibility of tonal components is higher for driving conditions at lower speeds. In contrast, an increase in speed results in an increase of the masking tire-road and wind noise components [83]. Therefore, the sounds that are recorded at higher speeds often sound more similar. For the *fl* condition, the same speed ranges as for the *ru* condition were selected.

73

The 20 sounds were divided as follows:

- 6 full-load run-up sounds from 0 km/h
- 7 full-load run-up sounds from 30 km/h
- 7 full-load run-up sounds from 50 km/h

For the *fl* condition, the numbers of sounds for the different speed ranges were chosen more equally because the full-load acceleration ability between the vehicles is considerably different, which resulted in different final speeds at the end of each stimulus. Furthermore, the audibility of tonal components was observed to be much higher with more pronounced level amplifications during structural resonances for sounds during full-load acceleration.

For the *cd* condition, the 60 sounds were divided as follows:

- 20 coast-down sounds from 30 km/h
- 20 coast-down sounds from 50 km/h
- 20 sounds from 70 km/h

The audibility of tonal components during coast-down depends on the usage of energy recuperation. For sounds, where the gearshift was shifted to the neutral gear so that no recuperation was used, the audibility of tonal components was markedly reduced. In total, 19 of 60 *cd* sounds were recorded in the neutral gear to circumvent recuperation. In contrast, for *cd* sounds with activated recuperation, tonal components were much more acoustically prominent.

5.1.2 Results

The following section shows the results of the 20 listeners, who participated in the magnitude estimation and pairwise comparison experiments.

Figure 25 shows the mean values and interindividual standard deviations for the sounds of the *ru* condition for both variables pleasantness (dark blue) and MOTC (yellow) for the different initial speeds 0 km/h (left panel), 30 km/h (center panel) and 50 km/h (right panel). Average pleasantness data range from 17.9 (little pleasant) to 74.6 (pleasant). Average MOTC data range from 20.0 (very little tonal) to 72.3 (tonal). On average, the stimuli with an initial speed of 0 km/h were rated more pleasant (mean pleasantness 64.8) than those with an initial speed of 30 km/h (48.8) and 50 km/h (36.5). Regarding the variable MOTC, on average, the stimuli with an initial speed of 0 km/h were rated more tonal (57.9) than those with an initial speed of 30 km/h (40.8) and 50 km/h (28.7).

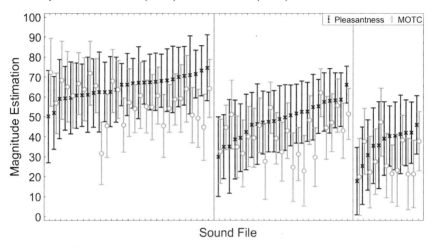

Figure 25. Mean values and standard deviations of the pleasantness (blue crosses) and MOTC (yellow circles) for the recorded electric vehicle interior sounds (abscissa) of the ru condition for the initial speeds 0 km/h (left panel), 30 km/h (center panel) and 50 km/h (right panel).

Figure 26 shows the mean values and interindividual standard deviations for the sounds of the *fl* condition for both variables pleasantness and MOTC. The initial speeds are the same as for the *ru* condition: 0 km/h (left panel), 30 km/h (center panel), 50 km/h (right panel). For the *fl* condition, average pleasantness data range from 28.3 (little pleasant) to 63.3 (pleasant). Average MOTC data range from 41.9 (moderately tonal) to 76.4 (tonal). On average, the *fl* sounds with an initial speed of 0 km/h were rated more pleasant (51.2) than those with an initial speed of 30 km/h (40.3) and 50 km/h (35.5). The MOTC also depended on the vehicle speed: Sounds with an initial speed of 0 km/h were rated more tonal (67.6) than those with an initial speed of 30 km/h (62.3) and 50 km/h (57.1).

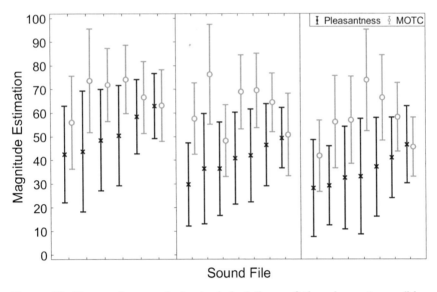

Figure 26. Mean values and standard deviations of the pleasantness (blue crosses) and MOTC (yellow circles) for the recorded electric vehicle interior sounds (abscissa) of the fl condition for the initial speeds 0 km/h (left panel), 30 km/h (center panel) and 50 km/h (right panel).

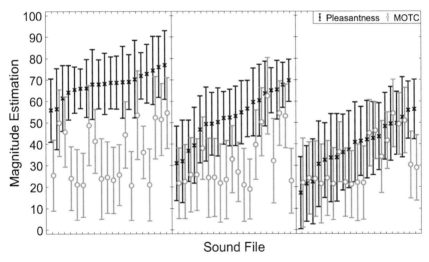

Figure 27. Mean values and standard deviations of the pleasantness (blue crosses) and MOTC (yellow circles) for the recorded electric vehicle interior sounds (abscissa) of the cd condition for the initial speeds 30 km/h (left panel), 50 km/h (center panel) and 70 km/h (right panel).

Figure 27 shows the mean values and standard deviations for the sounds of the *cd* condition for the pleasantness and MOTC with the initial speeds of 30 km/h (left panel), 50 km/h (center panel) and 70 km/h (right panel). Average pleasantness data for the *cd* condition range from 17.3 (very little pleasant) to 76.9 (pleasant). Average MOTC data range from 19.0 (very little tonal) to 72.4 (tonal). On average, the sounds of the *cd* condition with an initial speed of 30 km/h (67.8) were rated more pleasant than those with an initial speed of 50 km/h (52.8) and 70 km/h (39.4). For the *cd* condition, the vehicle speed only had a marginal influence on the perceived MOTC: Sounds with an initial speed of 30 km/h were rated slightly more tonal (35.4) than those with an initial speed of 50 km/h (32.1) and 70 km/h (32.1).

Similarly, to the experiments with artificial sounds, for each driving condition, a subset was chosen to conduct a complete pairwise comparison experiment for both variables pleasantness and MOTC. To limit the time consumption, twelve sounds were selected for each driving condition, which, according to equation (2), results in 66 comparisons. The sounds of each subset were chosen based on the following criteria:

- Discriminability of sounds with respect to the variables pleasantness and MOTC
- Large dynamic range with respect to both variables
- Equal representation of different vehicle speed ranges
- Choice of different vehicle models

For the pairwise comparison subset, the BTL scaled pleasantness values range from 0.009 to 0.242 (factor 28.1) for the *ru* condition, from 0.027 to 0.199 (factor 7.4) for the *fl* condition and from 0.004 to 0.308 (factor 74.9) for the *cd* condition. The correlation coefficients between the average pleasantness values of the magnitude estimation experiment and the BTL scaled data from the pairwise comparison experiment are 0.98 (*ru*), 0.95 (*fl*) and 0.99 (*cd*) and show that the data from both experiments agree with each other.

The dynamic ranges of the BTL scaled MOTC are from 0.013 to 0.235 (factor 13.3) for the *ru* condition, from 0.026 to 0.166 (factor 6.5) for the *fl* condition and from 0.013 to 0.239 (factor 18.7) for the *cd* condition. The correlation coefficients between the average MOTC of the magnitude estimation experiment and the BTL scaled MOTC are 0.98 (*ru*), 0.93 (*fl*) and

0.98 (*cd*) and show that also for the MOTC the BTL scaled data confirm the data from the magnitude estimation despite their high interindividual standard deviations.

5.2 Hybrid Driving Condition

5.2.1 Stimuli

For the hybrid vehicle interior sounds, a similar strategy was used for stimuli generation. The stimuli were selected from four different vehicles. The hybrid vehicle interior sounds were selected under the condition that the combustion engine had to be switched on at least once during the signal duration of five seconds. The driving conditions and speed ranges were chosen in the same way as for the pure-electric vehicles with the difference that due to the low number of available recordings there was no sub-categorization of different speed ranges for the *cd* condition. The low number of available recordings for the *cd* condition results from the mechanism that during coast-down the combustion engine of hybrid vehicles is switched off during most of the recordings and the electric motor is used as a generator for energy recuperation. In total, 53 sounds were evaluated, 33 for the *ru* condition, 12 for the *fl* condition and 8 for the *cd* condition. For the *ru* condition, the sounds were divided into the different speed ranges as follows:

- 19 run-up sounds from 0 km/h
- 10 run-up sounds from 30 km/h
- 4 run-up sounds from 50 km/h

The lower the speed of the hybrid vehicle and the lower the acceleration, the higher the audibility of the tonal components. The internal combustion engine emits a broadband masking noise [84], which, for low speeds, is even more dominant than the tire-road and wind noise [6]. Therefore, the focus was set to sounds, which are recorded at lower driving speeds. For the *fl* condition, strong influences of the tonal sound components of the combustion engine have been observed. Especially for higher vehicle speeds, the components from the electric motor are barely audible. The 12 sounds of the *fl* condition were separated into the different speed ranges as follows:

- 5 full-load run-up sounds from 0 km/h
- 4 full-load run-up sounds from 30 km/h
- 3 full-load run-up sounds from 50 km/h

5.2.2 Results

The following section shows the results of the 20 listeners, who participated in both the magnitude estimation and the complete pairwise comparison experiments.

Figure 28 shows the mean values and interindividual standard deviations of the pleasantness (dark blue crosses) and MOTC (yellow circles) for the *ru* condition. The panels are ordered by the initial speed from left to right: 0 km/h (left panel), 30 km/h (center panel), and 50 km/h (right panel). Average pleasantness data range from 31.7 (little pleasant) to 74.3 (pleasant). Average MOTC data range from 30.4 (little tonal) to 65.4 (tonal). On average, the sounds with an initial speed of 0 km/h were perceived as more pleasant (62.0) than those with an initial speed of 30 km/h (53.6) and 50 km/h (49.8). The MOTC did not differ much for the stimuli with different initial speeds: Sounds with an initial speed of 0 km/h were perceived as slightly less tonal (44.2) than those with an initial speed of 30 km/h (49.8) and 50 km/h (49.0).

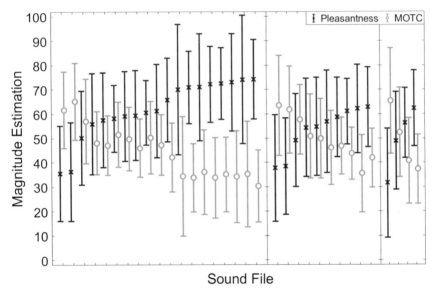

Figure 28. Mean values and interindividual standard deviations for the recorded hybrid vehicle interior sounds (abscissa) of the ru condition for the variables pleasantness (blue crosses) and MOTC (yellow circles). The left panel shows the ru sounds with an initial speed of 0 km/h, the center one those with an initial speed of 30 km/h and the right one those with an initial speed of 50 km/h.

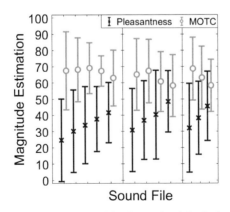

Figure 29. Mean values and interindividual standard deviations for the recorded hybrid vehicle interior sounds (abscissa) for the variables pleasantness (blue crosses) and MOTC (yellow circles). The left panel shows the sounds of the fl condition with an initial speed of 0 km/h, the center one those with an initial speed of 30 km/h and the right one those with an initial speed of 50 km/h.

Figure 29 shows the average pleasantness and MOTC values for the *fl* condition and the interindividual standard deviations. The data are ordered according to the initial speed in the same way as for the *ru* condition. The range of the values for the *fl* condition is relatively small, with values from 24.7 (little pleasant) to 45.5 (moderately pleasant) for the pleasantness and from 58.3 (moderately tonal) to 69.1 (tonal) for the MOTC. For the *fl* condition, both variables did not depend much on the initial speed: Stimuli with an initial speed of 0 km/h were perceived as slightly less pleasant (33.7) than those with an initial speed of 30 km/h (39.2) and 50 km/h (38.7). Regarding the MOTC, sounds with an initial speed of 0 km/h were perceived as slightly more tonal (67.0) than sounds with an initial speed of 30 km/h (62.9) and 50 km/h (63.4).

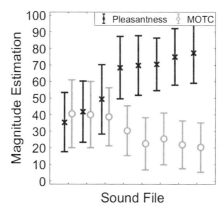

Figure 30. Mean values and interindividual standard deviations for the pleasantness (blue crosses) and MOTC (yellow circles) for the sounds of the cd condition (abscissa).

Figure 30 shows the mean values and standard deviations for the variables pleasantness and MOTC for the *cd* condition. Due to the low amount of available stimuli for hybrid vehicles where the combustion engine was switched on, the data were not subcategorized into different speed ranges. Average pleasantness values range from 35.4 (little pleasant) to 77.2 (pleasant). Average MOTC values range from 20.2 (little tonal) to 40.5 (moderately tonal).

Also for the hybrid vehicle stimuli, a pairwise comparison experiment was conducted for both variables to confirm the results of the magnitude estimation and to estimate the relative pleasantness and MOTC of both variables on a ratio scale. For the *ru* condition, 12 sounds were chosen according to the criteria described in section 5.1.2, which results in 66 comparisons. For the conditions *fl* and *cd*, all stimuli of the magnitude estimation experiment were included in the pairwise comparison experiment. The BTL scaled pleasantness range from 0.018 to 0.16 (factor 8.9) for the *ru* condition, from 0.021 to 0.18 (factor 8.6) for the *fl* condition and from 0.013 to 0.277 (factor 22) for the *cd* condition. The correlation coefficients of 0.98 (*ru*), 0.97 (*fl*) and 0.99 (*cd*) confirm the results from the magnitude estimation experiment.

The BTL scaled MOTC data range from 0.039 to 0.217 (factor 5.5) for the *ru* condition. For the *fl* condition, the data range from 0.059 to 0.116 (factor 2). For the *cd* condition, the MOTC data range from 0.071 to 0.288 (factor 4.1). The correlation coefficients for the three conditions are 0.96 (*ru*), 0.84 (*fl*) and 0.94 (*cd*). The lower correlation for the *fl* condition results from the lower dynamic range of the *fl* condition of both the magnitude estimation and the BTL scaled MOTC data.

5.3 Discussion

The results for the pure-electric vehicle interior sounds show a clear influence of the sound pressure level on the pleasantness. The sound pressure level is mainly determined by the background noise. The sounds of the *ru* and *fl* conditions with an initial speed of 0 km/h and of the *cd* condition with an initial speed of 30 km/h have a considerably quieter background noise due to the lower level of the tire-road and wind noise components than the sounds with higher initial speeds. Thus, the perceived pleasantness of these sounds is higher than for the sounds with higher initial speeds, which is in agreement with the results from the acoustic optimization experiment with synthetic sounds described in section 4.4. In that experiment, the maximum background noise was varied in two steps with a difference of 10 dB SPL in between. The sounds with the higher maximum background noise level were perceived as significantly less pleasant than those with the lower maximum background noise level, showing that the intake reduction of tire-road and wind noise into the vehicle interior is a primary goal also for vehicles with electrified drives.

The influence of the MOTC on the pleasantness of the pure-electric vehicle interior sounds is ambiguous. For the *ru* condition, the rank correlation according to Kendall [85] between both variables is 0.42, showing that, in general, sounds with a higher MOTC were perceived as more pleasant. However, when comparing the sounds of each initial speed condition with each other, this trend is not visible, indicating that not the MOTC is responsible for the higher pleasantness but the lower background noise level of the sounds with a higher MOTC. For the *fl* sounds, this relationship is less pronounced (rank correlation coefficient 0.16), which can be explained by two different reasons: Firstly, the dynamic range of both variables, especially the MOTC, is considerably lower than of the *ru* sounds. Secondly, the acceleration ability of the vehicles is considerably different. While the upper class vehicles accelerate the fastest (involving more pronounced and higher-frequency tonal components), these vehicles are also better insulated. For the *cd* condition, the perceived MOTC primarily depended on, whether recuperation was used or not. Sounds, where recuperation was used, were perceived as much more tonal (38.2) than sounds recorded in neutral gear with deactivated recuperation (22.5), while the pleasantness seems to be affected mainly by other factors, such as the level, which is shown by the pleasantness differences when comparing the three initial speed conditions.

The implications for the hybrid vehicle interior sounds are similar to the pure-electric vehicle interior sounds. However, due to the influence of the combustion engine, which, especially for lower vehicle speeds, dominates the impression of the sound, the dynamic range is considerably lower than that of the pure-electric vehicles.

6 Experiments with Augmented Sounds

To increase the data set for the development of a pleasantness model and to investigate the influence of the different sound components, experiments with augmented sounds were conducted. Augmented sounds are recorded sounds from the vehicle interior with spectro-temporal manipulations. Therefore, the results of the sound separation and allocation algorithm [2] were used. Sections 6.1 and 6.2 describe the experiments, where the levels of the electric motor orders and inverter components were varied. Sections 6.3 and 6.4 investigate, whether artificially generated subharmonics and noise could be used to increase the pleasantness of the vehicle interior sound.

6.1 Variation of Electric Motor Order Levels

6.1.1 Stimuli

The extracted electric motor orders were used either to increase the level or to remove a certain component. Therefore, 9 original stimuli (4 run-up, 2 full-load run-up, 3 coast-down) from four different vehicles were selected to manipulate the sound pressure level of each electric motor order (abbreviated as *EO*, followed by the respective order number), which could be binaurally extracted.

A preliminary sound analysis was conducted to determine the suitable stimuli and the appropriate spacing for the level manipulations of each extracted electric motor order. The sound analysis revealed that an increase by either 6, 12 or 24 dB SPL is suitable to ensure that the stimuli are sufficiently discriminable, which is indicated by the indices *+6 dB*, *+12 dB* and *+24 dB*. The maximum increase of 24 dB SPL was motivated by the fact that most of the orders were audible when their level was increased by 24 dB SPL. The preliminary sound analysis further revealed that for most of the orders, a stepwise reduction of the sound pressure level of each order does not ensure sufficient discriminability of the sounds. Therefore, in addition to the individual stepwise amplification of each electric motor order, each order was removed individually, which is indicated by the index *off*. To quantify the influence of the manipulation of each order, the original sound and a residual sound were presented. The residual sound *res* of each original stimulus was generated by removing all extractable electric motor orders from the original stimulus, including those, which could be only extracted monaurally.

The above-mentioned variations, in total, added up to 155 stimuli. To investigate the interdependencies between the variables pleasantness and MOTC, both variables were acquired during the experiment. As the previous experiments revealed a high agreement between the data from the magnitude estimation and those from the pairwise comparison experiment and the high time consumption of the magnitude estimation experiment (4-5 hours per listener for both variables), a complete pairwise comparison experiment has not been conducted.

6.1.2 Results

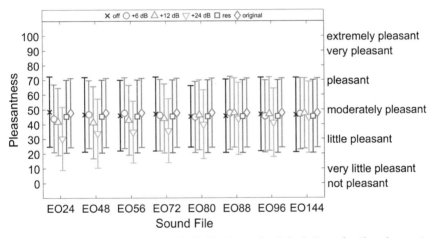

Figure 31. Mean values and interindividual standard deviations for the pleasantness of the sounds, where the level of each extracted electric motor order of a pure-electric vehicle in a run-up condition was separately manipulated. The data are grouped into groups of six data points, where the first four data points refer to the level manipulations of the order indicated on the abscissa. The blue crosses show the pleasantness of the sound, where the respective order was removed. The purple circles, the yellow upwards pointing triangles and the green downwards pointing triangles refer to the sounds, where the respective order level was increased by 6, 12 and 24 dB SPL. The grey squares show the pleasantness values of the residual sound, where all orders were removed, while the red diamonds show those of the original sound.

Figure 31 shows a representative example of the mean pleasantness values and their interindividual standard deviations of all level variations of all electric motor orders which could be binaurally extracted from a recorded full-load run-up sound with an initial speed of 50 km/h. The complete pleasantness results of the experiment are shown in Appendix A. The analysis revealed that in general, the pleasantness decreased when the level of a certain order was increased. The pleasantness decrease was higher if the level of a certain order was increased. In addition, the order number highly influenced the pleasantness: The lower the order number the higher the influence of the level increase on the pleasantness. In contrast, the removal of each individual order (*off*) and all orders (*res*) only marginally influenced the pleasantness perception compared to the *original* stimulus.

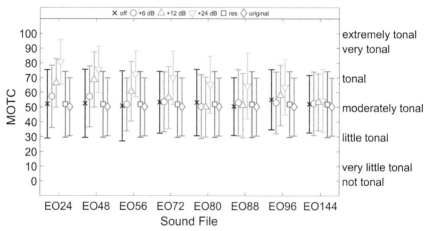

Figure 32. Same as Figure 31 but here for the variable MOTC.

Figure 32 shows the mean values of the perceived MOTC and their interindividual standard deviations for the same variations as shown in Figure 31. The complete MOTC results of the experiment are shown in Appendix A. The results show that the level increase has a higher impact on the MOTC than on the pleasantness: The higher the level increase of an order, the higher the perceived MOTC increased. Similar to the pleasantness perception, the effect was stronger for lower order numbers than for higher order numbers. As for the variable pleasantness, the removal of an order and the removal of all orders altogether only marginally influenced the MOTC perception.

6.2 Variation of Inverter Component Levels

6.2.1 Stimuli

To evaluate, which influence high-frequency components could have on the electric vehicle interior sound, the sound pressure level of the extracted inverter components was adjusted. In contrast to the electric motor orders, the sound pressure level of each inverter component was below the masked threshold in most of the cases. For the experiment, 8 original sounds (2 run-up, 2 full-load run-up, 4 coast-down) were used for the level manipulations of the extracted inverter components. Two kinds of inverter components were distinguished: While the inverter switching frequencies were abbreviated with *SF*, followed by the frequency, the inverter harmonics were abbreviated by the underlying switching frequency, followed by the subscripted order number the switching frequency is interacting with. For each binaurally extractable inverter component, either switching frequency or harmonic, the sound pressure level was adjusted individually. Due to the reason, that for most of the inverter components, the difference of the component level to the masked threshold was high, the level of the inverter orders was increased by up to 36 dB SPL.

To ensure, that the level of most of the inverter components is above the masked threshold for at least one variation, the level was increased by either 6, 12, 24 or 36 dB SPL, indicated by the indices *+6 dB*, *+12 dB*, *+24 dB* or *+36 dB*. Like in the previous experiment, the removal of each individual component was indicated by the index *off*, while the index *res* indicated, that all inverter components were removed, also those, which could only be extracted monaurally. As before, the *original* sound was included as a reference to quantify the influence of the level manipulations. The original stimuli were chosen in a way that the level manipulations of the components resulted in an audible difference at least for the largest level increase. A maximum increase of 36 dB SPL of each component is still below the sound pressure limit for the total sound, which was approved by the ethical committee.

All the above-mentioned manipulations added up to 324 variations. As in the previous experiment, both variables pleasantness and MOTC were evaluated in a magnitude estimation experiment. The conduction of a pairwise comparison subset was omitted in order to not further increase the experimental time of 6-7 hours per listener and because in the previous experiments the results from the magnitude estimation experiment and the BTL scaled results from the pairwise comparison experiment were highly consistent.

6.2.2 Results

Twenty listeners participated in the experiment. Figure 33 shows the mean pleasantness values and the interindividual standard deviations of a representative original sound, which was recorded in a coast-down condition with an initial speed of 50 km/h. The complete pleasantness data of the experiment are shown in Appendix B. The abscissa shows the adjusted components, which are ordered by their switching frequency. From the sound, four inverter switching frequencies, namely 5 kHz, 10 kHz, 15 kHz and 20 kHz, denoted as *SF5kHz*, *SF10kHz*, *SF15kHz* and *SF20kHz*, could be binaurally extracted. Furthermore, two lower-side inverter harmonics of the 5 kHz switching frequency, denoted as $SF5kHz_{-36}$ and $SF5kHz_{-28}$, were detected.

The results reveal that a level increase of each individual inverter component led to a decrease in the perceived pleasantness. The higher the frequency of the adjusted component, the stronger the effect on the pleasantness decrease. For the higher-frequency components *SF15kHz* and *SF20kHz*, only an increase of 36 dB SPL considerably affected the pleasantness. As for the previous experiment, the experimental results are strongly linked to the masked threshold of each individual component. If the sound pressure level after the level adjustment was still below the masked threshold, the adjustment did not have a sizable effect on the pleasantness. The experiment also revealed that for nearly all sounds, the effect of the removal of either each individual (*off*) or all inverter components (*res*) was neglectable compared to the *original* sound, since they were not audible at their original level in the vehicle interior.

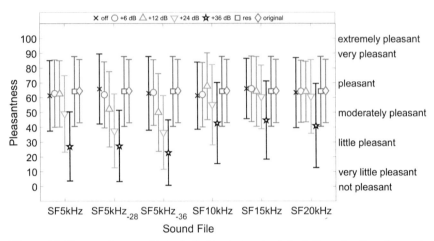

Figure 33. Mean values and interindividual standard deviations for the pleasantness of the sounds, where the level of each extracted inverter component of a pure-electric vehicle in a coast-down condition with an initial speed of 50 km/h was separately manipulated. The data are grouped into groups of seven data points, where the first five data points refer to the level manipulations of the inverter component indicated on the abscissa. The blue crosses show the pleasantness of the sound, where the respective inverter component was removed. The purple circles, the yellow upwards pointing triangles, the green downwards pointing triangles and the black stars refer to the sounds, where the respective component level was increased by 6, 12, 24 and 36 dB SPL. The grey squares show the pleasantness values of the residual sound, where all inverter components were removed, while the red diamonds show those of the original sound.

Figure 34 shows the mean values and standard deviations of the perceived MOTC. The complete MOTC data of the experiment are shown in Appendix B. As for the pleasantness, there were no remarkable differences between the *original* sound, the residual sound *res*, and the sounds with the removed component *off*. Only for the sounds, where the level of the two lowest-frequency components $SF5kHz_{-36}$ and $SF5kHz_{-28}$ was adjusted, the MOTC considerably increased, if their level was increased by 12 dB SPL. For the sounds, where the level of higher-frequency components was increased, the increases only took effect, if the sound pressure level was increased by 24 dB SPL (*SF5kHz*) or 36 dB SPL (*SF5kHz, SF10kHz, SF15kHz* and *SF20kHz*).

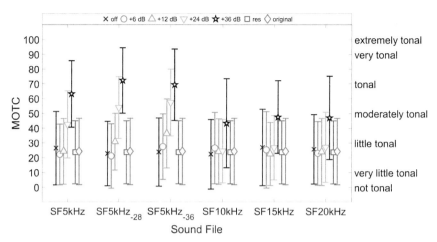

Figure 34. Same as Figure 33 but here for the variable MOTC.

In general, the experiment revealed that for most of the vehicles, the inverter components at their original level are not audible in the vehicle interior and can therefore be neglected for sound improvement measures.

6.3 Application of Subharmonics to Recorded Sounds

The following section describes the experiment, where artificially generated subharmonics were added to recorded vehicle interior sounds. Adding subharmonics could lower the perceived pitch of a tonal component by shifting the fundamental towards a lower frequency. That experiment aims to investigate whether subharmonics, which have been proven suitable to increase the pleasantness of synthetic vehicle interior sounds (see also section 4.4), are also suitable to increase the pleasantness of recorded vehicle interior sounds.

6.3.1 Stimuli

The experiment of the application of subharmonics could be regarded as an extension of the previously conducted experiment of applying subharmonics to synthesized sounds as described in section 4.4. Therefore, in total, seven original stimuli were selected. From these 7 stimuli, 2 were recorded in a run-up condition (1 pure-electric, 1 hybrid), 2 in a full-load run-up condition (1 pure-electric, 1 hybrid) and 3 in a coast-down condition (all pure-electric). In the following, the first part of the label of the variations of each original sound is the driving condition (*ru*, *fl* or *cd*), followed by the suffix _*Hyb* for the variations of the original sounds, which were recorded in a situation, where the combustion engine was switched on. These sounds were augmented with subharmonics in a similar way to the previous experiment. Therefore, the pitch of one dominant extracted electric motor or gear order was shifted by modifying the time scale of the extracted order using a phase vocoder [86, 87], followed by resampling to reduce the perceived pitch of the sound. The underlying order is the last (subscripted) part of the sound label, denoted as *EO*, followed by the respective order number. The result of this algorithm is a synthetic electric motor order with the same level over time as the original order. The resulting frequency content over time of the synthetic order has a constant factor to the one of the original order. With that algorithm, spectra of subharmonics consisting of synthetic motor orders like in the experiment on synthetic sounds could be created.

The following section describes all variations for the spectrum of tonal content (symbols in the result figures in parentheses). From the experiment with synthetic sounds, the spectra of tonal content _*II* and _*III* were used in the experiment on recorded sounds, because sounds with these spectra were perceived as most pleasant. To simplify the identification, in the following, the spectrum _*II*, which contains four subharmonics, where the frequency is halved and the level is increased by 6 dB towards lower frequency components, is denoted as spectrum _Oct_{+6dB} (yellow open upwards pointing triangles). Spectrum _*III*, which contains the two lowest-frequency components of spectrum _*II*, is denoted as spectrum _Oct_2l_{+6dB} (yellow open downwards pointing triangles). Due to the high levels of the tire-road and wind noise components of the original sounds, only spectra with a level increase towards lower-frequency subharmonics were used to augment the stimuli.

In addition to the spectra with an increase of 6 dB towards lower-frequency subharmonics, the same spectra with an increase of 12 dB were generated, denoted as _Oct_{+12dB} (yellow filled upwards pointing triangles) and _Oct_2l_{+12dB} (yellow filled downwards pointing triangles). The spectra with a level increase of 12 dB caused a large increase in the perceived loudness. The spectra of the tire-road and wind noise components had a low-pass characteristic so the lower-frequency components were masked more strongly than the higher-frequency ones. Considering this effect, additional spectra were generated, which contain the highest-frequency subharmonics of spectrum $\_Oct_{+12dB}$. Depending on the number of highest-frequency subharmonics (either three, two or one), the spectra were denoted as $\_Oct\_3_{+12dB}$ (yellow filled squares), $\_Oct\_2_{+12dB}$ (yellow filled stars) and $\_Oct\_1_{+12dB}$ (yellow filled diamonds).

In the previous experiment with synthetic sounds, the spectra with odd-numbered subharmonics $\_IV$ and $\_V$ were additionally used because they imitate the spectrum of a closed-pipe musical instrument [47]. Spectrum $\_IV$ contained two subharmonics with relative frequencies of one fifth and three fifths and spectrum $\_V$ contained three subharmonics with relative frequencies of one, three and five sevenths relative to the frequency of the underlying order. Both spectra had in common, that the level of each subharmonic was equal to the one of the underlying order. Therefore, the phenomenon, that the subharmonics were inaudible when they were applied to some recorded sounds, was also noticed for those spectra. Thus, those spectra were not applied to recorded sounds. Instead, both spectra were altered in a way, that the level increased by either 6 or 12 dB towards lower-frequency subharmonics. The resulting variations of spectrum $\_IV$ are denoted as $\_Odd\_2_{+6dB}$ (green open circles) and $\_Odd\_2_{+12dB}$ (green filled circles), the variations of spectrum $\_V$ as $\_Odd\_3_{+6dB}$ (red open right-pointing triangles) and $\_Odd\_3_{+12dB}$ (red filled right-pointing triangles). Additionally, the original sound without any subharmonics was presented as a reference, which is denoted as $\_none$ (blue crosses). The spectrum of subharmonics was the only parameter, which was varied in this experiment. In total, twelve variations of each original stimulus were presented, resulting in 84 stimuli for the whole experiment. For all stimuli, both variables pleasantness and MOTC were evaluated in a magnitude estimation experiment.

6.3.2 Results

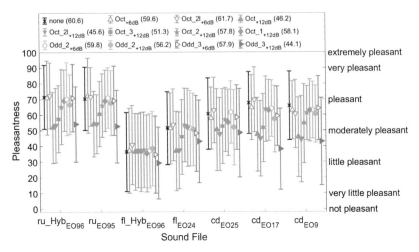

Figure 35. Mean values and interindividual standard deviations for the pleasant-ness. The data are grouped into sets of twelve data points, which show the variations of the spectrum of tonal content of each original sound. The blue crosses show the original sound. The yellow data points indicate the data of the sounds, where octave-spaced subharmonics were added. The green data points show the sounds with spectra of tonal content, which contained two subharmonics, while the red data points show those with spectra containing three subharmonics. The numbers in parentheses denote the mean values of the pleasantness of all stimuli with that spectrum of tonal content.

Figure 35 shows the mean pleasantness values and interindividual standard deviations of the 20 listeners. The average pleasantness data range from 29.3 (fl_Hyb_{EO96} with spectrum $\_Odd\_3_{+12dB}$) to 72.4 (ru_Hyb_{EO96} with spectrum Oct_2l_{+6dB}) with high interindividual standard deviations. The data show that for most of the sounds, small pleasantness improvements were achievable, while for one sound (cd_{EO9}), the pleasantness was lower for all spectra of tonal content, which contained subharmonics. The largest pleasantness improvements, on average, were achievable using spectrum $\_Oct\_2l_{+6dB}$ (61.7). Comparing the spectra, which only differ in their level (open vs. filled symbols), shows, that in general, a level increase of 6 dB per subharmonic was preferred over a level increase of 12 dB per subharmonic. The pleasantness data of the sounds, where the highest-frequency subharmonics were kept, i.e. $\_Oct\_3_{+12dB}$, $\_Oct\_2_{+12dB}$ and $\_Oct\_1_{+12dB}$, revealed that using a lower number of subharmonics with a higher level increase towards lower frequency is not a suitable approach. The sounds

94

with the odd-numbered spectra were, on average, also not perceived as more pleasant than the sounds without any subharmonics. While the spectra with two odd-numbered subharmonics _Odd_2_{+6dB} (59.8), _Odd_2_{+12dB} (56.2) and the sounds with the spectra with three odd-numbered subharmonics _Odd_3_{+6dB} (58.1) were perceived as only slightly less pleasant compared to the sounds without any subharmonics (60.6), the sounds with the spectrum _Odd_3_{+12dB}, were, on average, perceived as least pleasant (44.1).

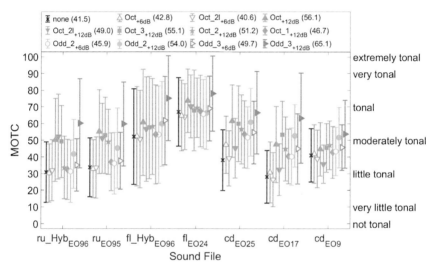

Figure 36. Same as Figure 35 but here for the variable MOTC.

Figure 36 shows the mean values for the MOTC for the same sounds. Mean MOTC values range from 26.3 (cd_{EO25} with spectrum _Oct_2l_{+6dB}) to 77.9 (fl_{EO24} with spectrum _Odd_3_{+12dB}). The data show that, on average, for the sounds with all spectra with subharmonics except _Oct_2l_{+6dB}, the perceived MOTC increased, while for the sounds with spectrum _Oct_2l_{+6dB}, the MOTC marginally decreased. In general, sounds with a higher number and level of subharmonics were perceived as more tonal than those with spectra with fewer subharmonics and lower levels. As an example, the sounds with spectrum _Oct_{+12dB} were perceived as much more tonal than those with the spectrum _Oct_{+6dB}. When removing the lower-frequency subharmonics from spectrum _Oct_{+12dB}, which results in

95

the spectra $\_Oct\_3_{+12dB}$, $\_Oct\_2_{+12dB}$ and $\_Oct\_1_{+12dB}$, the MOTC decreased as the number of subharmonics decreased. This relation also holds true when comparing the sounds with spectra containing odd-numbered subharmonics with each other: Sounds with three subharmonics were perceived as more tonal than those with two subharmonics. Sounds with spectra with a level increase of 12 dB towards lower-frequency subharmonics were perceived as more tonal than those with subharmonics with a level increase of 6 dB.

6.4 Variation of Tire-Road and Wind Noise Levels

The perceived pleasantness and MOTC strongly depend on the audibility of tonal components, mainly originating from the electric motor and the gearbox. An increase in the background noise level reduces the audibility of those components but also increases the total loudness of the sound. To investigate the influence of the tire-road and wind noise components, an experiment was conducted, where the sound pressure levels of both components were separately altered.

6.4.1 Stimuli

The previous experiments revealed that the sound pressure level of the background noise highly influenced both variables pleasantness and MOTC. Therefore, the experiment was conducted on real recordings from the vehicle interior, where the algorithm according to Fröhlingsdorf and Pischinger, 2022 [2] was used to separate the background noise, which mainly consists of the tire-road and wind noise components. For the experiment, 6 different original sounds were used (2 run-up, 2 full-load run-up, 2 coast-down), where the levels of the tire-road noise (tl) and the wind noise (wl) were adjusted individually. Both levels were either reduced by 12 dB SPL (indicated by $-12dB$), kept original ($0dB$) or increased by 12 dB SPL ($+12dB$). The combination of both level variations resulted in nine variations for each original stimulus. In total, for each driving condition, the variables pleasantness and MOTC of 18 stimuli were evaluated in a magnitude estimation experiment.

6.4.2 Results

Twenty listeners participated in the experiment. Figure 37 shows the mean pleasantness values and interindividual standard deviations for the six original sounds. The data are grouped into sets of nine data points for each original sound, where the label on the abscissa denotes the driving condition ru, fl and cd, followed by a subscripted index (1 or 2) to distinguish between the two original stimuli of each driving condition. Each set is subgrouped into subsets of three data points, which are colored green, yellow and red, to show the tire-road noise level of the respective stimulus. The green data points indicate a reduction of the tire-road noise level by 12 dB SPL (tl_{-12dB}), while the yellow data points indicate the original level (tl_{0dB}) and the red data points indicate an increase of the tire-road noise level by 12 dB SPL (tl_{+12dB}). In each subgroup, the respective wind noise level is shown by the marker symbol. The circles denote a level reduction of the wind noise by 12 dB SPL (wl_{-12dB}), the triangles denote the original wind noise level (wl_{0dB}) and the squares denote a level increase of the wind noise by 12 dB SPL (wl_{+12dB}). Within the legend, a horizontal line separates the adjustment of both levels.

The data reveal that both the tire-road and wind noise levels have a recognizable influence on perceived pleasantness. The sounds, where both noise levels were reduced by 12 dB SPL were, on average, perceived as more pleasant than those with the original levels and those, where the levels were increased. The pleasantness differences between the sounds with decreased and original noise levels were substantially smaller than between those with original and increased levels. As an example, compared to the sounds with the combination $tl_{0dB} \mid wl_{0dB}$, where the tire-road and wind noise levels were kept original, the pleasantness only increased from 56.0 to 58.4, when the tire-road noise level was reduced by 12 dB SPL, but decreased to 36.2 when the tire-road noise level was increased by 12 dB SPL. The influence of varying the wind noise was smaller: When keeping the tire-road noise level original, a reduction of the wind noise level by 12 dB SPL led, on average, to a marginal pleasantness increase from 56.0 to 56.8, while a level increase by 12 dB SPL caused a substantial pleasantness decrease from 56.0 to 41.7.

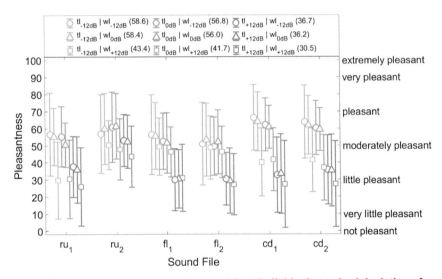

Figure 37. Mean pleasantness values and interindividual standard deviations for the sounds with manipulated background noise levels. The data are grouped into sets of nine data points, where each set contains all variations of one original stimulus. The scale ticks on the abscissa denote the two original stimuli of the run-up (ru), full-load run-up (fl) and coast-down (cd) driving conditions. Each set is subgrouped into three subsets, where the first subset (green) indicates the sounds, where the tire-road noise level was reduced by 12 dB SPL (tl-12dB). The second subgroup indicates the sounds with the original tire-road noise level (tl0dB). The third subgroup indicates the sounds, where the tire-road noise level was increased by 12 dB SPL (tl+12dB). Within each subgroup, the symbols denote the wind noise level. The circles indicate a reduction of 12 dB SPL (wl-12dB), while the triangles indicate the original wind noise level (wl0dB) and the squares indicate the sounds, where the wind noise level was increased by 12 dB SPL (wl+12dB). The numbers in parentheses show the mean value of all stimuli with the respective combination of tire-road and wind noise level adjustments.

Figure 38 shows the mean values and interindividual standard deviations for the MOTC. The data show that increasing both the levels of the tire-road and wind noise causes a substantial reduction of the perceived MOTC, while the influence is smaller when reducing the levels. Similar to the pleasantness, the tire-road noise variations influenced the perceived MOTC more than the wind-noise variations. Compared to the sounds with the original noise levels $tl_{0dB} \mid wl_{0dB}$, a reduction of the tire-road noise by 12 dB SPL caused, on average, a MOTC increase from 54.4 to 57.4, while an increase by 12 dB SPL caused an average MOTC reduction from 54.4 to 32.4. When the tire-road noise level was kept original, reducing the wind noise level only resulted in a MOTC increase from 54.4 to 56.4, while an increase reduced the average MOTC from 54.4 to 36.6.

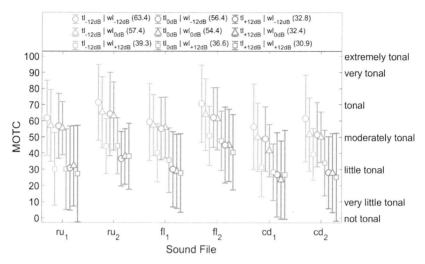

Figure 38. Same as Figure 37 but here for the variable MOTC.

6.5 Discussion

Within this chapter, the influence of single components on the variables pleasantness and MOTC was evaluated. For the electric motor orders, it was shown that the lower-frequency electric motor orders have a stronger influence on the pleasantness since their level is usually higher than that of the higher-frequency electric motor orders, which are partially or completely masked by the background noise. The decrease in pleasantness, if the level of the respective order was increased could be therefore explained by the increase of the MOTC. The results are in line with those of the study by Lennström et al., 2013 [42]. They concluded that an increase in the prominence ratio of the 36th and 72nd electric motor orders increases the perceived annoyance of the sound, while for an increase in the prominence ratio of the sixth electric motor order, a prominence ratio up to 5 dB does not induce higher annoyance ratings. The results of the electric motor order variation experiment are further in line with the results of Lennström et al., 2019 [88]. They showed that only for the higher-frequency tonal components with frequencies of 600 and 900 Hz, the pleasantness reduced with an increase of the detectability due to an increase in the prominence ratio. In contrast, tonal components with a frequency of 200 Hz were usually not detected and tonal components with a frequency of 400 Hz were detected, but an increase in the prominence ratio did not decrease the perceived pleasantness. Referring to Figure 31 and Figure 32 of the electric motor order level variation experiment, where the 24th order corresponds to a frequency range between 1200 and 3000 Hz within the signal duration of five seconds, it could be therefore concluded that a decrease of the perceived pitch as e.g. of the 24th order by adding subharmonics might be a suitable approach to increase the perceived pleasantness.

The frequencies of the inverter components are considerably higher than those of the electric motor. Referring to Figure 33 and Figure 34 of the inverter component level variation experiment, the results show that between the original sound and the sound without the specific inverter component, no difference is visible. Therefore, it could be concluded that the inverter components are usually not relevant in the vehicle interior due to their low level and that acoustic measures do not have to be conducted for inverter components.

The application of subharmonics to recorded sound is an extension of the preliminary experiment described in section 4.4 and the previously conducted studies of Gwak et al., 2014 [47] and Sun et al., 2018 [48]. The comparison with the acoustic optimization experiment with synthetic sounds, described in section 4.3, exposes certain differences. While in the previous experiment, on average, all spectra of subharmonics led to a pleasantness increase, in the current experiment, when including all sounds, only the addition of spectrum $\_Oct\_2l_{+6dB}$ led to a small increase (60.6 vs. 61.7) compared to the sounds without subharmonics. The differences could arise because of multiple reasons: In general, the background noise levels of the recorded sounds were higher than those of the synthetic sounds. Furthermore, the sound pressure levels of the underlying orders of the recorded sounds were generally lower than for the underlying orders of the synthetic sounds. Both effects accounted for a lower potential pleasantness improvement when adding subharmonics.

The third reason is likely the dependency of the subharmonics' level on the level of the underlying order. In contrast to the previous experiment, the extracted order's level was not constant over time but included audible level fluctuations. Therefore, also the subharmonics incorporate audible level fluctuations. Some listeners reported in the interview, which was conducted after the experiment, that the audible level fluctuations negatively influence the pleasantness of the stimuli. The experiment on generalized synthetic sounds (section 4.3) revealed that both the number and intensity of the structural resonances have a significant negative influence on pleasantness. The audible level fluctuations could therefore overweigh the pleasantness improvement resulting from the reduction of the perceived pitch.

The results also point out that the pitch reduction approach is only suitable if the original sound contains tonal components, which elicit a relatively high-frequency pitch. As an example, the largest pleasantness improvement was achieved on the original sound fl_Hyb_{EO96} using the spectrum of tonal content $\_Oct\_2l_{+6dB}$. The underlying 96th order had a frequency range from 2.9 kHz to 4.1 kHz over a signal duration of five seconds. Therefore, the application of spectrum $\_Oct\_2l_{+6dB}$ achieved a frequency shift down to a frequency range between 180 and 260 Hz for the lowest-frequency subharmonic, while the loudness according to the dynamic loudness model (DLM) of Chalupper and Fastl [82] (see section 7.1), on average, only increased from 11.901 to 11.911 sone (left channel) and from 11.206 to 11.211 sone (right channel). As a counterexample, the extracted order with the lowest frequency of the original sound cd_{EO9} was the ninth order with a much lower frequency range from 530 to 210 Hz, so that the lowest-frequency subharmonic of the most-preferred spectrum $\_Odd\_3_{+6dB}$ ranged from 76 to 30 Hz. The loudness when applying that most-preferred spectrum increased from 5.12 to 5.57 sone (left channel) and from 4.73 to 5.02 sone (right channel).

The experiment concludes that adding subharmonics might be a suitable technique to improve the pleasantness of electric vehicle interior sounds. Therefore, the masking effects of the tire-road and wind noise components (see section 6.4) should be considered. The level of the subharmonics over time should further not be directly linked to the level of the underlying order as this relation might result in audible amplitude modulations, which will negatively affect the perceived pleasantness.

Similarly, to the preliminary experiment, the tire-road and wind noise levels showed a large influence on both variables. While a decrease of both levels revealed to only have a relatively small potential to increase the pleasantness, an increase of the level largely reduces the pleasantness. Similarly, the MOTC increase due to a noise level decrease of 12 dB was considerably lower than the MOTC decrease when each noise level was increased by 12 dB. Therefore, it could be resumed, that the MOTC decrease does not compensate for the loudness increase due to the noise level increase. When discussing the results of the experiment on noise level variations in the light of the experiment on subharmonics, a further level decrease of the tire-road and wind noise components does not only appear useful for a level decrease but further increases the audibility of the applied subharmonics. As both noise components have low-pass-shaped spectra, their level increase will simplify the application of subharmonics to lower the perceived pitch caused by tonal components from the electric motor and gearbox.

7 Development of a Pleasantness Assessment Model

The results of all previously conducted experiments with recorded and augmented sounds, which were described in chapters 5 and 6, were used to estimate a model to enable a pleasantness prediction for vehicles with electrified drives without time-consuming and costly jury evaluations. In total, 702 sounds were used to estimate the pleasantness model. Among these 702 sounds, 198 sounds were recorded sounds, while 504 were sounds from the data augmentation experiments, where either the levels of certain sound components were varied or synthetic sound components were added. As the dynamic perceptual aspects are highly relevant for the perception of the target variable pleasantness, the time-variant psychoacoustic parameters were mapped to the average single values of the pleasantness for each sound. Therefore, the conception strategy for the pleasantness model was as follows:

Firstly, several time-varying parameters, such as loudness, sharpness and tonality were calculated. They were used as input data for the model estimation. Secondly, the data were preprocessed in a way that they are suitable for the following estimation of the pleasantness model. Thirdly, a suitable model architecture was chosen, which considered the time-varying aspects of the psychoacoustic parameters and the dependencies between the time steps of the parameters. The last step was the model estimation and validation: An iterative optimization algorithm was used to calculate the optimal model parameters with to minimize the error between the predicted pleasantness and the average pleasantness from the hearing experiments.

7.1 Calculation of Potential Predictors

Several parameters were considered as potential predictors. As the input data had two channels, the data were separately calculated for both channels and averaged after they were downsampled to a common sampling rate. The most commonly investigated influence parameter on pleasantness is loudness. Several studies showed that the loudness has a negative effect on the pleasantness perception of vehicle interior sounds [39, 41, 89]. Therefore, it was used as a key parameter for the estimation of the pleasantness model. The previously conducted experiments showed that the dynamic aspects of the loudness are relevant for the pleasantness perception so a dynamic model considering these aspects has to be selected. The dynamic loudness model (DLM) according to Chalupper and Fastl, 2002 [82], which is based on the loudness model according to Zwicker, 1960 [90], consists of several sequential stages. A high-pass filter, which accounts for the absolute threshold of the first critical band, is followed by a critical band filter bank. The following envelope extraction stage extracts the temporal envelopes from the outputs of the filters. The following transmission factor represents the transmission between free field and the human hearing system. The resulting excitation is then transformed into a specific loudness before spectral and temporal masking effects are considered. The last stages are the spectral summation and the temporal integration: The spectral summation stage is used to calculate the total loudness from the specific loudness values of each critical band. The temporal integration stage simulates duration effects on the resulting loudness. Two different outputs of the model were used for the following model estimation stage: The resulting time-varying loudness $N(t)$, which was used as a predictor, is a vector, whose values denote the loudness at each point in time in sone. A pure tone with a frequency of 1 kHz and a sound pressure level of 40 dB SPL is defined to have a loudness of one sone so that the loudness of other sounds could be related to the loudness of that reference sound. The other outputs are the specific time-varying loudness values $N'(z, t)$ for each critical band z, which were used to calculate the sharpness over time, described in the following paragraph.

The second potential predictor is the time-varying sharpness $S(t)$, which was calculated according to the German standard DIN 45692 [91]. The sharpness depends on the spectral envelope of the signal [8]. The higher the high-frequency spectral content of a signal, the higher the sharpness value. As a reference, a narrow-band noise with a critical bandwidth of one Bark and a center frequency of 1000 Hz is defined to have a sharpness of one acum, so that the sharpness of other sounds could be determined in relation to that reference sound. As inputs, the specific time-varying loudness values $N'(z, t)$ according to Chalupper and Fastl, 2002 [82] were used to calculate the sharpness over time. They were weighted with the weight $g(z)$ according to the following formula:

$$S(t) = k * \frac{\int_{z=0\,Bark}^{z=24\,Bark} N'(z,t) * g(z) * z * dz}{\int_{z=0\,Bark}^{z=24\,Bark} N'(z,t) * dz} \tag{6}$$

with $g(z) = 1$ for $z \leq 15.8\,Bark$ and $g(z) = 0.15 * e^{0.42(z-15.8)} + 0.85$. The normalization constant k is used to adjust the formula in a way that calculating the sharpness of the reference narrow-band noise results in a value of one acum. The resulting time-varying sharpness is used as a predictor for the pleasantness model.

The third parameter, which was considered for the model estimation, is the tonality according to ECMA-418-2 [51], which is equivalent to the previously used term magnitude of tonal content (MOTC). The tonality according to that standard uses a calculation method to calculate the specific loudness, which is separated into the loudness of the tonal components and the loudness of the background noise using an autocorrelation function. For each band, the tonal loudness is estimated by evaluating the spectrum of the autocorrelation function. A subsequent noise reduction stage compensates for the overestimation of the specific loudness in the previous step. To determine the time-dependent specific tonality, the signal-to-noise ratio across all bands is calculated. The last step of the model is to calculate the time-dependent tonality. It is determined by taking the maximum of all time-dependent specific tonality values. In a previous study, it was proved, that the resulting tonality is a good predictor for the MOTC for the time-varying synthetic vehicle interior sounds described in section 4.3, see also Doleschal et al., 2020 [53]. The time-varying tonality according to ECMA-418-2 was therefore used as a predictor for the estimation of the pleasantness model.

A further considered parameter was the roughness, which could be particularly relevant for vehicles with hybrid vehicles, where the combustion engine is switched on [92, 93]. Therefore, the roughness according to Daniel and Weber, 1997 [94] was considered. After the model transforms the input signal into 200-ms long frames, which are weighted by a Blackman window, the spectra for all windows are calculated and the frequency components are transformed into the excitation pattern. Using a critical-band filter bank of 47 channels with a bandwidth of one Bark and an overlap of 0.5 Bark each, the specific excitation spectrum of each channel is obtained. For each channel, a generalized modulation depth is calculated by dividing the root-mean-square value of the weighted excitation envelope by its constant component value. As the perceived roughness depends on the carrier frequency, the generalized modulation depths are weighted by multiplying each of them with a weighting function. To avoid an overestimation in the case of noise signals due to random modulations, the cross-correlation coefficients of each band with their neighboring ones are multiplied by the product of the modulation depth and the weighting factor. The squared product results in the specific roughness, which is then summed up over all 47 channels and multiplied with a calibration factor. The calibration factor ensures that for a reference sound, a 100% amplitude-modulated 1 kHz tone with a modulation frequency of 70 Hz and a sound pressure level of 60 dB SPL, the model outputs a roughness value of one asper. For the model estimation, the roughness over time was considered a potential predictor.

7.2 Data Preprocessing

For the subsequent model estimation, the data had to be preprocessed. The predictors loudness, sharpness, tonality and roughness had different sampling rates. Therefore, in the first step, they were downsampled to a common sampling rate. A sampling rate of 100 Hz (equivalent to one data point every 10 ms) is lower than the output sampling rates of all potential predictors and is sufficiently high for inputs of sound quality prediction models regarding vehicle sounds [95]. In a second step, the predictors of the left and right channels were averaged, so that for each predictor, a vector with a length of 500 samples for a signal with a duration of five seconds described the respective variable over time.

To ensure the robustness of the model [96, 97], the whole data set was randomly split into three subsets: training set, validation set and test set. The training set was used to train the parameters of the prediction model. The validation set was used to tune the hyperparameters of the model (see also section 7.4). The samples of the test set were used to evaluate the predictive performance of the model [97] after the training process finished. If the percentage of the training set is chosen too low, the training algorithm might underfit the relation between the input parameters and the pleasantness. If the percentage of the training set is chosen too high, the model will poorly predict the pleasantness of sounds, which were not used for training [98], which is called overfitting. Commonly, a percentage between 50 and 80 percent [96, 97, 99, 100] of the data set is used for training the model. Due to the relatively low number of samples (in total 702), the percentage for the training set was chosen conservatively, which resulted in a value of 60% (421 samples). The rest of the data were equally split (20 % each) into the validation set (140 samples) and the test set (141 samples).

The last step of the preprocessing was the normalization of the data. Normalization should be applied because the parameters loudness, sharpness, tonality and roughness have different value ranges [101] so that without normalization the training algorithms will not perform effectively, which results in high training durations and poor prediction results [102]. The input data were normalized using a zero-center normalization. The zero-center normalization was carried out in a way that the mean value of each variable (e.g. the mean loudness across all stimuli) was subtracted from each value of that variable so that after normalization the mean value across all stimuli was zero for each variable [103]. The resulting feature vectors, consisting of 500 normalized time-dependent values for each stimulus were used for the following model estimation.

7.3 Choice of Model Architecture

For the pleasantness model, several model types and architectures were considered. The simplest model architecture is linear regression. Its advantages are the straightforward interpretability of the resulting equations and the short calculation times using a least squares approximation. The main drawback is the assumption of a linear relationship between the input and output variables, which, in the case of the pleasantness model could not be assumed [39] due to several reasons: Firstly, the input parameters loudness, sharpness, tonality and roughness could not take negative values. Secondly, the target variable pleasantness could not take values outside the desired range between 0 and 100. A more sophisticated approach to overcome this problem is the usage of suitable transformations of the input and target variables to avoid the pleasantness could take impermissible values outside the desired value range [104]. Doleschal et al., 2019 [22] showed, that a log-linear approach is suitable to predict the pleasantness of the interior sounds of vehicles with combustion engines.

All equation-based regression approaches have the drawback in common, that a preliminary assumption of a functional relationship and its structure has to be assumed before model training, which is often not known in advance. The assumption of an unsuitable equation structure could lead to suboptimal results.

The assumption of a functional relationship could be avoided using machine-learning algorithms. Verhey et al., 2021 [105] showed that a Gaussian Process Regression (GPR) based model is suitable to predict the magnitude of the sensation booming. The model is based on observations, which are regarded as part of a multivariate Gaussian distribution so that the booming magnitude of new data points could be predicted based on the value pairs, which are near them. An advantage of the model is further, that the accuracy of each predicted data point could be estimated, depending on the proximity of the neighboring data points [106]. As model inputs, the results of the dynamic loudness model according to Chalupper and Fastl, 2002 [82] in combination with the modulation filter bank model according to Dau et al., 1997 [107] were used. The model revealed a high agreement for both the training and the test data, which is shown by correlation coefficients between the mean values from the hearing experiment and the model predictions of r=0.99 for the training set and of r=0.94 for the test set.

Another commonly used approach is using neural networks. Neural networks are defined as a directed graph consisting of neurons and edges, where the edges are the connections between the neurons. The neurons could be classified into three different types: input neurons, hidden neurons and output neurons. Input neurons are those, which receive the input parameters, and output neurons those, that output the output parameters. Those in between are defined as hidden neurons because they are only connected with other neurons and do not interact with the environment. The edges between the neurons are weighted. During the learning process, either new edges between nodes are created, existing ones removed or the weights are updated. The central mechanism of the learning process is the backpropagation algorithm so that the error between the predicted and the previously provided input data is used to update the network. A gradient descent algorithm commonly determines the direction, in which the data are updated. The gradient descent algorithm determines, which parameter update reduces the error between the predicted and the provided data at most [108].

Several types of neural networks have been previously described: Neural networks, which do not include feedback loops, are defined as feedforward networks; those, which do are defined as recurrent neural networks. Recurrent neural networks involve feedback loops and could save previous states, which is advantageous in the case of sequential inputs. One of the main drawbacks of recurrent neural networks is the vanishing gradient problem. When the error is propagated backwards over multiple layers, the gradient tends to become smaller in every layer. This means, that its effect on updating the weights becomes smaller and smaller, which causes long learning times and suboptimal results. One way to overcome that drawback is the usage of long short-term memory networks [109].

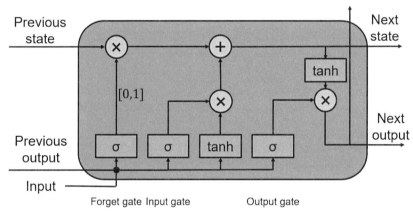

Figure 39. Overview of the structure of a LSTM cell.

Especially for vehicles with electrified drives, the dynamic aspects of sounds are particularly relevant. As an example, in a full-load run-up driving condition, the tonal electric motor and gear orders are prominent at low vehicle speeds at the beginning of a stimulus, which results in high values for the parameters sharpness and tonality at the signal onset. Similarly, for coast-down sounds, the onset particularly influences the pleasantness perception, as their loudness is highest at the beginning of the signal [108]. One solution, which enables the consideration of dependencies between the time steps of sequential input parameters [109] without facing the vanishing gradient problem, is using long short-term neural networks (LSTM).

The structure of an LSTM cell is shown in Figure 39. An LSTM cell contains three different types of gates: a forget gate, an input gate and an output gate. The forget gate decides, which information will be kept and which will be discarded. The forget gate is realized by a sigmoid function, where a value of zero means, that all information should be discarded and a value of one that all information of the previous state should be stored. The input gate, which is also realized by a sigmoid function, decides which information is saved in the cell state. The information to be saved in the cell state, is processed by the tanh function and is multiplied with the results of the sigmoid function of the input gate and added to the information, which passed through the forget gate. The sum of the previous information and the current information will then serve as the previous state of the subsequent cell. The output gate decides which parts of the cell state are going to be the output. Like the other gates, it is based on a sigmoid function, resulting in values between zero and one. The cell state goes through the tanh function and the product of the output gate values and the tanh function is the next output and serves as the previous output for the subsequent LSTM cell [109].

A subtype of an LSTM is the so-called bidirectional LSTM (biLSTM), which considers dependencies between the time steps of the inputs in both directions, so that the predictive accuracy of the model can often be improved compared to a unidirectional LSTM [110]. As dependencies of the psychoacoustic parameters in both directions are particularly relevant for predicting the pleasantness of vehicle interior sounds, a biLSTM-based model structure is chosen.

The sequential input parameters could be either regressed to a sequential output (sequence-to-sequence regression) [111, 112] or to a single value (sequence-to-one regression) [113, 114]. In the case of this study, the pleasantness was acquired as a single value, because the task of rating the time-varying pleasantness by synchronously moving the slider during stimulus playback was too difficult for the listeners due to the highly dynamic changes of the investigated stimuli.

Neural networks have many practical applications in acoustics: Ma et al., 2017 [39] pointed out that a three-layer feedforward neural network, which used the parameters A-weighted sound pressure level, loudness, fluctuation strength, tonality, roughness, sharpness and articulation index is suitable to predict the pleasantness and four other variables of pure-electric vehicle interior sounds, evaluated in a semantic differential experiment. Steinbach and Altinsoy, 2019 [41] created several neural networks for the prediction of the annoyance of synthesized pass-by sounds and compared their predictive performance to the performance of a linear regression approach. Except for the articulation index, they used the same psychoacoustic parameters as Ma et al., 2017 [39]. The comparison revealed that the predictive performance of the linear regression approach is lower than the one of the optimal neural network, which could be explained by two reasons: Firstly, their study revealed, that, except for the loudness, the relation between the input parameters and the annoyance could not be assumed as linear. In addition, the input parameters (e.g. A-weighted sound pressure level and loudness) were partially correlated, which decreases the performance of an equation-based regression approach. The feedforward network approaches in the aforementioned studies all have in common that they use a single value for each input parameter (e.g. mean value or the maximum) and that dynamic aspects, which particularly have to be considered for electric vehicle interior sounds, could not be properly addressed using that kind of approach.

A common application for LSTMs in acoustics is the prediction of sensations. Hadzalic, 2018 [113] showed that a long short-term memory neural network is suitable to predict the perceived annoyance of aircraft interior sounds, which were evaluated by listeners in a psychoacoustic experiment. In that study, the normalized time signals of the stimuli were used as inputs for an LSTM with the annoyance as output. The study compared the performance of the LSTM with the one of a spectrogram-based convolutional neural network (CNN). They pointed out, that the LSTM is generally able to predict pleasantness; however, the predictive performance of the CNN was better. The result could be discussed under the aspect that using time series of psychoacoustic predictors might be a better-suited approach for the prediction of sensations.

Therefore, as mentioned before, the current study used the parameters loudness, sharpness, tonality and roughness as potential predictors and the mean pleasantness values from the psychoacoustic experiments on recorded and augmented sounds as targets. The architecture of the

111

biLSTM, which consisted of six layers, is shown in Figure 40. For better visualization, the three biLSTM layers were aggregated into one block.

Figure 40. Structure of the neural network used for the pleasantness prediction, consisting of a sequence input layer, three bidirectional LSTM layers, a fully connected layer and an output layer. For a better visualization, the three LSTM layers were aggregated into one block.

The first layer was a sequence input layer. The sequence input layer defined the dimension of the input data matrix (number of currently selected predictors x 500) and preprocessed the data as described in section 7.2. Three biLSTM layers followed the sequence layer. The biLSTM layers model the dependencies between the sequential input parameters and the single-valued target parameter pleasantness. The number of biLSTM layers was determined iteratively. A higher number of biLSTM layers enables a higher accuracy when modelling complex relationships, but also increases the training time of the model. The next block describes a fully connected layer: The fully connected layer is designed to learn the functional relationship between the output of the last biLSTM layer and the provided target values [115]. The last layer of the model was the regression layer. It calculates the error between the target output from the experiments and the predicted output so that the output of the regression layer was used to determine the predictive accuracy of the model.

To avoid overfitting, the inclusion of dropout layers after each biLSTM layer was evaluated. Overfitting means, that the model predicts the training data well but fails to predict the test data and any further previously unknown data. The dropout layer randomly sets a certain percentage of the input elements to zero, the percentage of elements is called the dropout ratio [116]. According to Srivastava et al., 2014 [116], a dropout ratio between 20 and 50 percent is optimal. Therefore, the percentage was varied in steps of five percent and compared to the results without dropout. The results revealed that the usage of dropout layers did not further improve the prediction results of the model, so dropout layers have not been included into the final model.

7.4 Model Estimation and Validation

The following section describes the implementation process and the iterative tuning process of the model parameters. As mentioned in the previous section, neural networks do not require to assume a certain functional relationship between the input and output parameters. Instead, the structure of the model, including the number of biLSTM layers, the considered input parameters and the hyperparameters were optimized. Hyperparameters are parameters, which have to be set before model training. Even though they do not directly describe the relationship between the input and the target variables, their setting is often crucial for the accuracy of the resulting model [117].

The parameter variation was carried out in several steps. At first, reasonable starting values for the hyperparameters and the number of biLSTM layers were assumed. Afterwards, the number of biLSTM layers and the hyperparameters were optimized with respect to minimal root-mean-square errors (RMSE) for the training set and the test set. For the hyperparameter optimization process, the input variables loudness N, sharpness S and tonality T were considered, as it was likely that these parameters are most relevant for the pleasantness perception of vehicles with electrified drives (see also section 7.1). The following hyperparameters were considered:

- Number of hidden units
- Maximum number of training epochs
- Initial learn rate
- Learn rate drop period
- Learn rate drop factor

Hidden units store the information between time steps. The higher the number of hidden units, the more information between time steps will be remembered [118]. However, a large number of hidden units could make the model prone to overfitting [119]. Furthermore, a large number increases the complexity of the model, which results in high calculation times [118]. To reduce the possible number of values in the parameter optimization process, the number of hidden neurons was set equal for every biLSTM layer. To limit the calculation time in the beginning, the number of hidden units was set to a starting value of 12 and increased in the following hyperparameter optimization process. The analysis revealed that a value of 150 hidden neurons was the optimal value and that a further increase of the number did not result in a further improvement of the resulting model predictions, but in extremely high training times and instability issues.

For the same reason to limit the calculation time during the parameter optimization process, the starting value for the number of training epochs was set to six. In each epoch, the gradient descent algorithm takes one optimization step of the internal model parameters in order to decrease the error between the predicted values and the target data [120]. The value was increased in a stepwise procedure until a value of 24 training epochs was reached. After 24 training epochs, the error between the target data and the predicted data did not substantially reduce anymore.

The parameter learn rate determines the slope of the gradient descent algorithm [121]. Even though the accuracy of the final estimation does not depend much on the exact value of the learn rate [122], the training time increases if the learn rate is too low. However, in the case of choosing both the parameters learn rate and the number of maximum epochs too low, the model parameters do not converge to an optimal solution. In contrast, choosing a too high value for the learn rate will cause an unstable training process. Therefore, at the beginning of the training process, the learn rate should be higher than in the later training process. The starting value for the initial learn rate was set to a value of 0.1. The value of the learn rate drop period was set to 3 and of the learn rate drop factor to 0.5, which means, that every three epochs the learn rate was halved. The following optimization process revealed, that an initial learn rate of 0.1 and a learn rate drop factor of 0.5 in combination with a learn rate drop period of three epochs resulted in the smallest prediction errors for both the training and the test set.

With that optimal set of hyperparameters, the optimal combination of input parameters was determined. Therefore, a forward and backward selection was conducted. For the forward selection process, every variable is separately added to a model without any parameters. After adding each variable separately, the best one-parameter model is chosen as a basis for the next iteration step. As a quality criterion, the RMSE of the test set is selected to ensure the robustness of the final model. This procedure is repeated until no parameter addition improves the model any further or until all potential predictors are included in the model. The backward selection is conducted in the opposite way: The model with all input parameters is regarded as a basis. Every variable is separately removed and the best remaining model is used for the next iteration step.

Table 11. Table of the forward and backward selection of the model estimation process. The columns show the combination of input variables, the root-mean-square errors (RMSE) for the training and test sets and the correlation coefficients between the predicted data and the target data for both sets. The optimal models with one, two and three parameters are highlighted green.

Combination of Inputs	$RMSE_{Train}$	$RMSE_{Test}$	r_{Train}	r_{Test}
N	8.0	8.1	0.81	0.79
T	11.5	12.2	0.50	0.48
S	7.2	8.4	0.84	0.81
N, T	7.0	6.9	0.86	0.85
N, S	7.7	6.9	0.82	0.86
N, T, S	6.7	5.2	0.88	0.92

The results of the forward and backward selection are shown in Table 11. The roughness R was not related to the variable pleasantness and had a very low dynamic range, so it was excluded for further model estimation. It was shown, that the best one-parameter model only considered the loudness N with a training RMSE of $RMSE_{Train} = 8.0$, a test RMSE of $RMSE_{Test} = 8.1$ and correlation coefficients between the predicted and the experimental data of $r_{Train} = 0.81$ for the training set and of $r_{Test} = 0.79$ for the test set. The best two-parameter model was a model considering the loudness N and the tonality T with RMSEs of $RMSE_{Train} = 7.0$ and $RMSE_{Test} = 6.9$ and correlation coefficients of $r_{Train} = 0.86$ and $r_{Test} = 0.85$. The best three-parameter model was a model considering the loudness N, the tonality T and the sharpness S with $RMSE_{Train} = 6.7$, $RMSE_{Test} = 5.2$ and correlation coefficients of $r_{Train} = 0.88$ and $r_{Test} = 0.92$. The conduction of the backward selection revealed that a model containing only the parameters loudness and sharpness has a lower predictive accuracy than the three-parameter model. The three-parameter model has been chosen as the final model to predict the pleasantness of vehicles with electrified drives.

Considering the whole dataset (training set, validation set and test set), Figure 41 shows, that for both pure-electric (left panel) and hybrid vehicle interior sounds, the experimental data and the predicted data show a high correlation. For the pure-electric vehicle sounds, which include hybrid vehicle sounds with switched-off combustion engines, the RMSE had a value of 6.6 and the correlation coefficient between predicted and experimental data was 0.89. For the hybrid vehicle sounds, the RMSE was 6.9 and the correlation coefficient was 0.86. The results demonstrate, that the computed bidirectional long short-term memory network with the above-mentioned hyperparameter set, considering the loudness N, the tonality T and the sharpness S as input variables, is suitable to predict the pleasantness of vehicle interior sounds of both pure-electric and hybrid vehicles.

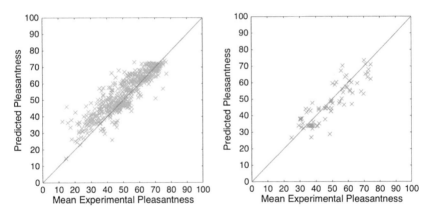

Figure 41. Predicted pleasantness in comparison to the mean experimental pleasantness for pure-electric (left panel) and hybrid (right panel) vehicles. The comparison reveals that for both sets, the predictions and the experimental results are highly correlated.

7.5 Temporal Analysis of Parameter Influences

To investigate, whether certain parts of the stimuli are more relevant than other parts and if generally small snippets could explain the pleasantness of a whole stimulus, a temporal analysis was conducted. In that temporal analysis, all one-parameter models and the total model were estimated using the same procedure as mentioned before, but using only a small temporal portion of the signal. Therefore, the stimuli were cut into snippets of 50 ms each and the model estimation process was run using the same model structure and the same set of hyperparameters as before. For each snippet, the RMSE was calculated and plotted as a data point in Figure 42. The figure demonstrates that for the one-parameter models with the parameter loudness and sharpness and for the total model, the beginning of the stimuli is slightly less relevant for the pleasantness than the rest of the stimuli. This result could be interpreted in a way that for the run-up and full-load run-up sounds both parameters are very low at signal onset and are not suitable as predictors for the pleasantness, while the tonality is a relative measure of the tonal to the noisy content of the signal. The analysis further showed that the test set RMSE of the total model using a 50-ms snippet is considerably higher than the one using the whole stimuli, indicating that not a certain portion of the stimuli accounts for the pleasantness and that the dynamic aspects of the input variables for the whole stimuli have to be considered to achieve an optimal prediction result.

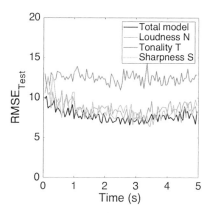

Figure 42. Root-mean-square errors of the three one-parameter models of the variables loudness (yellow), tonality (red) and sharpness (green) and the total model (blue) using all parameters. For the model estimation, a 50-ms-long snippet around the time value of each data point was used.

8 Summary and Conclusions

The study aimed to analyze the influence parameters on pleasantness, to estimate and validate a dynamic pleasantness assessment model and to give suggestions for active sound design measures in order to increase the pleasantness of the vehicle interior sound of vehicles with electrified drives. Therefore, several experiments with synthetic, recorded and augmented stimuli were conducted to determine the audibility of the separated sound components as well as the perceived pleasantness and the magnitude of tonal content (MOTC).

The data augmentation experiments using the separated sound components revealed that the tonal electric motors are relevant for the pleasantness perception, while the level of the inverter components in the vehicle interior is commonly below the masked threshold so that they have no or little influence on the pleasantness of the interior sound. The noisy tire-road and wind noise components strongly influence the pleasantness. The level reduction of the tire-road and wind noise is not only desired due to their large contribution to the overall loudness of the sounds but also due to the stronger masking of applied subharmonics, which could be introduced to lower the pitch elicited by high-frequency tonal components. The comparison between the pleasantness and MOTC revealed that in most cases, a higher MOTC leads to a lower pleasantness.

An exception is the application of subharmonics. In some cases, the addition of subharmonics increases the pleasantness of the vehicle interior sound, even though the perceived MOTC increases due to a higher number of tonal components. The increase was higher for the synthetic sounds than for the recorded sounds, because, for the synthetic sounds, the prominence of the electric motor orders was higher due to the higher order level and the lower background noise level, whereas the preferred spectra were the same for both the synthetic and recorded sounds. Thus, the integration of subharmonics as part of an active sound design concept should be considered. The generation of the sounds using a vocoder-based approach and the presentation via the infotainment system does not incur considerably higher costs. In contrast, the sound generation using a sound module, which is integrated into the power electronics of the electric motor and the presentation via the electric motor itself involves higher costs, but might increase the authenticity of the vehicle interior sound. At the same time, a level reduction of the tire-road and wind noise intake is still desired to decrease the overall level and to enable the audibility of the applied subharmonics.

The experiments revealed, that the psychoacoustic parameters loudness, sharpness and MOTC explain the pleasantness, which is confirmed by the subsequent model development. The work further exposed that a long short-term memory model (LSTM) approach is suitable to model the variable pleasantness while considering the dynamic changes of the psychoacoustic parameters over time, as this kind of model enables the consideration of long-term dependencies between the time steps of psychoacoustic parameters. A combination of the time-varying loudness, tonality and sharpness with a sampling rate of 100 data points per second was found to be appropriate to model the experimentally evaluated single-valued pleasantness. The RMSE values were 6.7 for the training set and 5.2 for the test set. The correlation coefficients between the predicted data and the mean values from the magnitude estimation experiment were 0.88 for the training set and 0.92 for the test set, showing that the assessment model is robust and enables the pleasantness prediction of new stimuli, which were not used for the training process of the model.

Together with the separation and allocation algorithm, the pleasantness assessment model has been implemented into a software tool with a graphical user interface, which allows the application of the model for users without a deeper understanding of the underlying mechanisms. In the future, this tool will further allow the simulation of sound design measures. Therefore, the mechanism of generating subharmonics will be implemented into the tool, so that the user could generate customized sounds to assess the impact of sound design measures on the interior noise before implementing them into the vehicle.

References

[1] S. Schneider, F. Doleschal, J. Hots, A. Oetjen, H. Rottengruber and J. Verhey, "Benchmark-Analyse verschiedener Tonhaltigkeits-modelle anhand von Luftschallmessungen an aktuellen Fahrzeug-Elektromotoren," in *Fortschritte der Akustik - DAGA 2020*, Hannover, Deutsche Gesellschaft für Akustik e. V., 2020, pp. 330-332.

[2] K. Fröhlingsdorf and S. Pischinger, "Automatisierte Analyse des Innengeräuschs von Elektrofahrzeugen," in *Fortschritte der Akustik - DAGA 2022*, Stuttgart, Deutsche Gesellschaft für Akustik e. V., 2022, pp. 433-436.

[3] N. C. Otto, R. Simpson and J. Wiederhold, "Electric Vehicle Sound Quality," *SAE Technical Paper 1999-01-1694*, 1999.

[4] Q. Kang, P. Gu, C. Gong and S. Zuo, "Test and Analysis of Electromagnetic Noise of an Electric Motor in a Pure Electric Car," *SAE Technical Paper 2019-01-1492*, 2019.

[5] J. F. Gieras, C. Wang and J. C. Lai, Noise of Polyphase Electric Motors, Boca Raton: Taylor & Francis, 2006.

[6] K. Engel, B. Snitil, K. Wolff, R. Handel, J. Krüger, F. Gauterin, M. Helfer and O. Brass, "Wesentliche Geräuschquellen im Fahrzeug und deren Charakterisierung," in *Sound-Engineering im Automobilbereich*, Berlin Heidelberg, Springer, 2010, pp. 205-316.

[7] H. Naunheimer, B. Bertsche, J. Ryborz, W. Novak and P. Fietkau, "Auslesung von Zahnrädern für Fahrzeuggetriebe," in *Fahrzeug-getriebe: Grundlagen, Auswahl, Auslegung und Konstruktion*, Berlin Heidelberg, Springer, 2019, pp. 275-306.

[8] H. Fastl and E. Zwicker, Psychoacoustics: Facts and models, Berlin Heidelberg: Springer, 2007.

[9] G. Rösel, P. Mönius, S. Spas and N. Daun, "Inverter- und Motoroptimierung mittels SiC-Technologie," *Motortechnische Zeitschrift 82 (1)*, pp. 54-59, 2021.

[10] W. C. Lo, C. C. Chan, Z. Q. Zhu, L. Xu, D. Howe and K. T. Chau, "Acoustic Noise Radiated by PWM-Controlled Induction Machine Drives," *IEEE Transactions on Industrial Electronics 47 (4)*, pp. 880-889, 2000.

[11] P. Zeller, "Fahrgeräusch," in *Handbuch Fahrzeugakustik*, Wiesbaden, Vieweg+Teubner, 2009, pp. 158-170.

[12] K. Iwao and I. Yamazaki, "A study on the mechanism of tire/road noise," *JSAE Review 17 (2)*, pp. 139-144, 1996.

[13] O. Bschorr, "Reduktion von Reifenlärm," *Automobil-Industrie 31 (6)*, 1986.

[14] C. Popp and M. Hintzsche, "Lärmarme Reifen und geräusch-mindernde Fahrbahnbeläge: Erkenntnisse – Maßnahmen – Konzepte," Landesanstalt für Umweltschutz Baden-Württemberg, Karlsruhe, 2004.

[15] Bundesamt für Umwelt BAFU - Abteilung Lärmbekämpfung und nichtionisierende Strahlung, "Der Einfluss des Reifens auf die Lärmbelastung des Strassenverkehrs," 2012.

[16] E.-U. Saemann, "Reifen-Fahrbahngeräusch," in *Handbuch Fahrzeugakustik*, Berlin Heidelberg, Springer, 2009, pp. 223-238.

[17] K.-U. Nam, D. I. Kang and H. L. Jeong, "Windgeräuschminderung am Unterboden durch akustische Holografie," *Automobiltechnische Zeitschrift 110 (6)*, pp. 480-488, 2008.

[18] M. Helfer, "Reifen-Fahrbahn-Geräusch und Umströmungsgeräusch von Kraftfahrzeugen," in *Fortschritte der Akustik - DAGA 2007*, Stuttgart, Deutsche Gesellschaft für Akustik e. V., 2007, pp. 743-744.

[19] G.-T. Badel, F. Doleschal and J. Verhey, "Spektro-temporale Geräuschmanipulationen als Grundlage zur Erforschung der Empfindungsgröße Wummern," in *Fortschritte der Akustik - DAGA 2020*, Hannover, Deutsche Gesellschaft für Akustik e. V., 2020, pp. 313-314.

[20] A. Heider, A. Schröder and C. Wagner, "Experimentelle Untersuchungen strömungsinduzierter Resonanzen überströmter Hohlräume," in *18. DGLR-Fachsymposium der STAB, 06. - 07. Nov. 2012*, Stuttgart, Deutsches Zentrum für Luft- und Raumfahrt e.V., 2012, pp. 194-195.

[21] C. Schumann, F. Doleschal, S. Pischinger and J. Verhey, "Separation, Allocation and Psychoacoustic Evaluation of Vehicle Interior Noise," *SAE Technical Paper 2019-01-1518*, 2019.

[22] F. Doleschal, C. Schumann, J. Verhey and S. Pischinger, "Trennung, Zuordnung und psychoakustische Bewertung von Fahrzeuginnengeräuschen," in *Fortschritte der Akustik - DAGA 2019*, Rostock, Deutsche Gesellschaft für Akustik e. V., 2019, pp. 137-140.

[23] C. Schumann, F. Doleschal, S. Pischinger and J. Verhey, "Entwicklung eines Analysewerkzeugs zur Erkennung und Bewertung von störenden Geräuschanteilen im Fahrzeuginnenraum," in *Fortschritte der Akustik - DAGA 2019*, Rostock, Deutsche Gesellschaft für Akustik e. V., 2019, pp. 639-642.

[24] S. Schneider, F. Doleschal, H. Rottengruber and J. Verhey, "Psychoakustische Bewertung verbrennungsmotorischer Geräusche," *Automobiltechnische Zeitschrift 124 (1)*, pp. 56-61, 2022.

[25] S. Schneider, J. Hots, T. Luft, H. Rottengruber, J. Verhey and H.-P. Rabl, "Entwicklung einer empirischen Formel zur Bewertung der Tickergeräuschanteile von Motorgeräuschen," *Fortschritte der Akustik - DAGA 2019*, pp. 643-646, 2019.

[26] F. Doleschal, J. Verhey, C. Schumann and S. Pischinger, "Effiziente Optimierung des Fahrzeuginnengeräuschs," *Automobiltechnische Zeitschrift 122 (1)*, pp. 76-81, 2020.

[27] T. Dau, D. Püschel and A. Kohlrausch, "A quantitative model of the 'effective' signal processing in the auditory system: I. Model structure," *The Journal of the Acoustical Society of America 99 (6)*, pp. 3615-3622, 1996.

[28] K. Fröhlingsdorf and S. Pischinger, "Motordrehzahlbestimmung aus dem Innengeräusch von Elektrofahrzeugen," in *Fortschritte der Akustik - DAGA 2022*, Stuttgart, Deutsche Gesellschaft für Akustik e. V., 2022, pp. 413-415.

[29] N. Aggarwal and W. C. Karl, "Line Detection in Images Through Regularized Hough Transform," *IEEE Transactions on Image Processing 15 (3)*, pp. 582-591, 2006.

[30] R. J. Samworth, "Optimal weighted nearest neighbour classifiers," *The Annals of Statistics 40 (5)*, pp. 2733-2763, 2012.

[31] R. A. Kirsch, "Computer determination of the constituent structure of biological images," *Computers and biomedical research 4 (3)*, pp. 315-328, 1971.

[32] N. Otsu, "A threshold selection method from gray-level histograms," *IEEE Transactions on Systems, Man, and Cybernetics 9 (1)*, pp. 62-66, 1979.

[33] D. De Klerk and A. Ossipov, "Operational transfer path analysis: theory, guidelines and tire noise application," *Mechanical Systems and Signal Processing 24 (7)*, pp. 1950-1962, 2010.

[34] R. Guski, "Psychological Methods for Evaluating Sound Quality and Assessing Acoustic Information," *Acta Acustica united with Acustica 83 (5)*, pp. 765-774, 1997.

[35] F. Aletta, J. Kang and Ö. Axelsson, "Soundscape descriptors and a conceptual framework for developing predictive soundscape models," *Landscape and Urban Planning 149*, pp. 65-74, 2016.

[36] A. Fenko, H. Schifferstein and P. Hekkert, "Noisy products: Does appearance matter?," *International Journal of Design 5 (3)*, pp. 77-87, 2011.

[37] DIN EN ISO 9000: Qualitätsmanagementsysteme - Grundlagen und Begriffe (ISO 9000:2015), Berlin: Beuth, 2015.

[38] H. Murata, H. Tanaka, H. Takada and Y. Ohsasa, "Sound Quality Evaluation of Passenger Vehicle Interior Noise," *SAE Technical Paper 1993-05-01*, 1993.

[39] C. Ma, C. Chen, Q. Liu, H. Gao, Q. Li, H. Gao and Y. Shen, "Sound Quality Evaluation of the Interior Noise of Pure Electric Vehicle Based on Neural Network Model," *IEEE Transactions on Industrial Electronics 64 (12)*, pp. 9442-9450, 2017.

[40] D. Garson, "Interpreting neural-network connection weights," *AI Expert 6 (7)*, pp. 47-51, 1991.

[41] L. Steinbach and E. Altinsoy, "Prediction of annoyance evaluations of electric vehicle noise by using artificial neural networks," *Applied Acoustics 145*, pp. 149-158, 2019.

[42] D. Lennström, T. Lindbom and A. Nykänen, "Prominence of tones in electric vehicle interior noise," in *International Congress and Exposition on Noise Control Engineering: 15/09/2013-18/09/2013*, Innsbruck, Österreichischer Arbeitsring für Lärmbekämpfung, 2013, pp. 508-515.

[43] H. Bao and I. Panahi, "Psychoacoustic Active Noise Control with ITU-R 468 Noise Weighting and its Sound Quality Analysis," in *2010 Annual International Conference of the IEEE Engineering in Medicine and Biology*, Buenos Aires, Institute of Electrical and Electronics Engineers, 2010, pp. 4323-4326.

[44] G. Eisele, P. Genender and K. Wolff, "Electric vehicle sound design - Just wishful thinking? Sounddesign von Elektrofahrzeugen - Ein Wunschgedanke?," FEV Motorentechnik GmbH, Aachen.

[45] Deutsches Institut für Normung e. V., "DIN 45681 Akustik - Bestimmung der Tonhaltigkeit von Geräuschen und Ermittlung eines Tonzuschlages für die Beurteilung von Geräuschimmissionen," Beuth, Berlin, 2003.

[46] International Organization of Standardization, "Acoustics — Description, measurement and assessment of environmental noise — Part 2: Determination of environmental noise levels," International Organization of Standardization, Genf, 2007.

[47] D. Y. Gwak, Y. Kiseop, Y. Seong and S. Lee, "Application of subharmonics for active sound design of electric vehicles," *The Journal of the Acoustical Society of America 136 (EL391),* pp. EL391-EL397, 2014.

[48] Y. Sun, Z. Yongji, H. Yifei and Y. Zhufang, "The reseach of improving the sound quality of electric vehicles by using subharmonics," *Proceedings of InterNoise/ASME NCAD Noise Control and Acoustics Division Conference - 2018 : presented at InterNoise/ASME 2018, August 26-29, 2018, Chicago, Illinois,* pp. 1468-1475, 2018.

[49] F. Doleschal and J. L. Verhey, "Pleasantness and magnitude of tonal content of electric vehicle interior sounds containing subharmonics," *Applied Acoustics 185,* p. 108442, 2022.

[50] S. He, J. Miller, V. Peddi, B. Omell and M. Gandham, "Active Masking of Tonal Noise using Motor-Based Acoustic Generator to Improve EV Sound Quality," *SAE Technical Paper 2021-01-1021,* 2021.

[51] "Psychoacoustic metrics for ITT equipment — Part 2 (models based on human perception)," Ecma International, Genf, 2020.

[52] R. Parncutt, Harmony: A psychoacoustical approach, Berlin Heidelberg: Springer, 2012.

[53] F. Doleschal, J. Hots, J. Verhey, A. Oetjen, S. Schneider and H. Rottengruber, "Tonality benchmark analysis for electric vehicle interior noise," in *INTER-NOISE and NOISE-CON Congress and Conference Proceedings, InterNoise 2020, Seoul, South Korea*, Seoul, Institute of Noise Control Engineering, 2020, pp. 2894-2902.

[54] M. Yu and J.-S. Park, "Development of Dash insulation with PU elastomer based sound insulation materials for increasing the sound insulation performances of electric vehicle noise derived from motor," in *INTER-NOISE and NOISE-CON Congress and Conference Proceedings*, Institute of Noise Control Engineering, 2021.

[55] M. Bodden and T. Belschner, "Principles of Active Sound Design for electric vehicles," *Proceedings of the Inter-Noise 2016 : 45th International Congress and Exposition on Noise Control Engineering : towards a quieter future : August 21-24, 2016*, pp. 1693-1697, 2016.

[56] S. Moon, S. Park, D. Park, M. Yun, K. Chang and D. Park, "Active Sound Design Development Based on the Harmonics of Main Order From Engine Sound," *Journal of the Audio Engineering Society 68 (7/8)*, pp. 532-544, 2020.

[57] M. Bodden and T. Belschner, "Comprehensive Automotive Active Sound Design part 1: electric and combustion vehicles," *INTER-NOISE and NOISE-CON Congress and Conference Proceedings*, p. Institute of Noise Control Engineering, 2014.

[58] L. Kinsler and A. Frey, Fundamentals of Acoustics, Hoboken: John Wiley & Sons, 1962.

[59] F. Doleschal, H. Rottengruber and J. L. Verhey, "Influence parameters on the perceived magnitude of tonal content of electric vehicle interior sounds," *Applied Acoustics 181*, p. 108155, 2021.

[60] H. Levitt, "Transformed up-down methods in psychoacoustics," *The Journal of the Acoustical Society of America 49 (2B)*, pp. 467-477, 1971.

[61] M. Nitschmann and J. L. Verhey, "Modulation cues influence binaural masking-level difference in masking-pattern experiments," *The Journal of the Acoustical Society of America 131 (3)*, pp. EL223-EL228, 2012.

[62] V. Hohmann, "Frequency analysis and synthesis using a Gammatone filterbank," *Acta Acustica united with Acustica 88 (3)*, pp. 433-442, 2002.

[63] B. R. Glasberg and B. C. J. Moore, "Derivation of auditory filter shapes from notched-noise data," *Hearing Research 47 (1-2)*, pp. 103-138, 1990.

[64] M. Nitschmann and J. Verhey, "Binaural notched-noise masking and auditory-filter shape," *The Journal of the Acoustical Society of America 133 (4)*, pp. 2262-2271, 2013.

[65] T. Dau, D. Püschel and A. Kohlrausch, "A quantitative model of the "effective" signal processing in the auditory system. II. Simulations and measurements," *The Journal of the Acoustical Society of America 99 (6)*, pp. 3623-3631, 1996.

[66] T. Brand and V. Hohmann, "An adaptive procedure for categorical loudness scaling," *Journal of the Acoustical Society of America 112 (4)*, pp. 1597-1604, 2002.

[67] R. T. Ross, "Optimum orders for the presentation of pairs in the method of paired comparisons," *Journal of Educational Psychology 25 (5)*, pp. 375-382, 1934.

[68] R. A. Bradley and M. E. Terry, "Rank Analysis of Incomplete Block Designs: I. The Method of Paired Comparisons," *Biometrika 41 (3/4)*, pp. 324-345, 1952.

[69] R. D. Luce, Individual choice behavior: A theoretical analysis, Mineola, New York: Dover Publications, 2005.

[70] F. Wickelmaier and C. Schmid, "A Matlab function to estimate choice model parameters from paired-comparison data," *Behavior Research Methods, Instruments, & Computers 36 (1)*, pp. 29-40, 2004.

[71] G. H. Jang and D. K. Lieu, "The effect of magnet geometry on electric motor vibration," *IEEE Transactions on Magnetics 27 (6)*, pp. 5202-5204, 1991.

[72] X. Zeng, J. Liette, S. Noll and R. Singh, "Analysis of Motor Vibration Isolation System with Focus on Mount Resonances for Application to Electric Vehicles," *SAE International Journal of Alternative Powertrains 4 (2)*, pp. 370-377, 2015.

[73] R. Sottek and W. Bray, "Application of a New Perceptually-Accurate Tonality Assessment Method," *SAE International Journal of Passenger Cars: Electronic and Electrical Systems 8 (2015-01-2282)*, pp. 462-469, 2015.

[74] S. A. White, "Design of a Digital Biquadratic Peaking or Notch Filter for Digital Audio Equalization," *Journal of the Audio Engineering Society 34 (6)*, pp. 479-483, 1986.

[75] M. Vormann, J. Verhey, V. Mellert and A. Schick, "Subjective Rating of Tonal Components in Noise with an Adaptive Procedure," *Contributions to Psychological Acoustics, Results of the 8th Oldenburg Symposium on Psychological Acoustics,* pp. 145-153, 2000.

[76] R. P. Hellman, "Perceived magnitude of two-tone-noise complexes: Loudness, annoyance, and noisiness," *The Journal of the Acoustical Society of America 77 (4)*, pp. 1497-1504, 1985.

[77] U. Landström, E. Åkerlund, A. Kjellberg and M. Tesarz, "Exposure levels, tonal components, and noise annoyance in working environments," *Environment International 21 (3)*, pp. 265-275, 1995.

[78] M. Vormann, J. L. Verhey, V. Mellert and A. Schick, "Ein adaptives Verfahren zur Bestimmung der subjektiven Tonhaltigkeit," *Fortschritte der Akustik - DAGA 2000,* pp. 304-305, 2000.

[79] R. P. Hellman, "Growth rate of loudness, annoyance, and noisiness as a function of tone location within the noise spectrum," *The Journal of the Acoustical Society of America 75 (1)*, pp. 209-218, 1984.

[80] H. Hansen and R. Weber, "The influence of tone length and S/N-ratio on the perception of tonal content: An Application of probabilistic choice models in car acoustics," *Acoustical Science and Technology 29 (2)*, pp. 156-166, 2008.

[81] H. Hansen, R. Weber and U. Letens, "Quantifying tonal phenomena in interior car sounds," *Proceedings Forum Acusticum 2005, Budapest, Hungary*, pp. 1725-1729, 2005.

[82] J. Chalupper and H. Fastl, "Dynamic Loudness Model (DLM) for Normal and Hearing-Impaired Listeners," *Acta Acustica united with Acustica 88 (3)*, pp. 378-386, 2002.

[83] G. Cerrato, "Automotive Sound Quality – powertrain, road and wind noise," *Sound and Vibration 43 (4)*, pp. 16-24, 2009.

[84] G. Lepoittevin, J. Horak and D. Caprioli, "The New Challenges of NVH Package for BEVs," *SAE Technical Paper 2019-01-1452*, 2019.

[85] M. Kendall, "A New Measure of Rank Correlation," *Biometrika 30 (1/2)*, pp. 81-93, 1938.

[86] J. Driedger and M. Müller, "A Review of Time-Scale Modification of Music Signals," *Applied Sciences 6 (2)*, p. 57, 2016.

[87] J. Driedger, Time-Scale Modification Algorithms for Music Audio Signals, Saarbrücken: Saarland University, 2011.

[88] D. Lennström, A. Gutierrez Meilan, A. Nykänen, J. Becker and R. Sottek, "Perception of tones below 1 kHz in electric vehicles," in *INTER-NOISE and NOISE-CON Congress and Conference Proceedings*, Madrid, Institute of Noise Control Engineering, 2019, pp. 3675-3685.

[89] Y. Wang, G. Shen and X. Yanfeng, "A sound quality model for objective synthesis evaluation of vehicle interior noise based on artificial neural network," *Mechanical Systems and Signal Processing 45 (1)*, pp. 255-266, 2014.

[90] E. Zwicker, "Ein Verfahren zur Berechnung der Lautstärke," *Acta Acustica united with Acustia 10 (4)*, pp. 304-308, 1960.

[91] Normenausschuss Akustik, Lärmminderung und Schwingungstechnik (NALS) im DIN und VDI, "DIN 45692: Messtechnische Simulation der Hörempfindung Schärfe," Beuth, Berlin, 2009.

[92] Y. Wang, G. Shen, H. Guo, X. L. Tang and T. Hamade, "Roughness modelling based on human auditory perception for sound quality evaluation of vehicle interior noise," *Journal of Sound and Vibration 332 (16)*, pp. 3893-3904, 2013.

[93] M. Pflüger and R. Höldrich, "Psychoacoustic Measurement of Roughness of Vehicle Interior Noise," in *37th Meeting of the Acoustical Society of America, 2nd Convention of the European Acoustics Association*, Berlin, European Acoustics Association, 1999.

[94] P. Daniel and R. Weber, "Psychoacoustical Roughness: Implementation of an Optimized Model," *Acta Acustica united with Acustica 83 (1)*, pp. 113-123, 1997.

[95] K. Liang and H. Zhao, "Automatic evaluation of internal combustion engine noise based on an auditory model," *Shock and Vibration 2019*, 2019.

[96] M. Coto-Jiménez, "Robustness of LSTM neural networks for the enhancement of spectral parameters in noisy speech signals," in *Mexican International Conference on Artificial Intelligence*, Cham, Springer, 2018, pp. 227-238.

[97] M. Sangiorgio and F. Dercole, "Robustness of LSTM neural networks for multi-step forecasting of chaotic time series," *Chaos, Solitons and Fractals - Nonlinear Science, and Nonequilibrium and Complex Phenomena 139*, p. 110045, 2020.

[98] H. Jabbar and R. Khan, "Methods to avoid over-fitting and under-fitting in supervised machine learning (comparative study)," *Computer Science, Communication and Instrumentation Devices,* pp. 163-172, 2015.

[99] Y. Duan, L. Yisheng and F.-Y. Wang, "Travel time prediction with LSTM neural network," in *Travel time prediction with LSTM neural network." 2016 IEEE 19th international conference on intelligent transportation systems (ITSC),* Rio de Janeiro, Institute of Electrical and Electronics Engineers, 2016, pp. 1053-1058.

[100] L. Prechelt, "PROBEN 1-a set of benchmarks and benchmarking rules for neural network training algorithms," Universität Karlsruhe, Karlsruhe, 1994.

[101] D. Borkin, "Impact of data normalization on classification model accuracy," *Research Papers Faculty of Materials Science and Technology Slovak University of Technology 27 (45),* pp. 79-84, 2019.

[102] J. Sola and J. Sevilla, "Importance of Input Data Normalization for the Application of Neural Networks to Complex Industrial Problems," *IEEE Transactions on Nuclear Science 44 (3) ,* pp. 1464-1468, 1997.

[103] S. Palaniswamy, "A Robust Pose & Illumination Invariant Emotion Recognition from Facial Images using Deep Learning for Human-Machine Interface," in *2019 4th International Conference on Computational Systems and Information Technology for Sustainable Solution (CSITSS),* Bengaluru, Institute of Electrical and Electronics Engineers, 2019, pp. 1-6.

[104] R. Taagepera, Making Social Sciences More Scientific: The Need for Predictive Models, Oxford: Oxford University Press, 2011.

[105] J. Verhey, G.-T. Badel and F. Doleschal, "Modellierung der Empfindung „Dröhnen" im Fahrzeuginnengeräusch," in *Fortschritte der Akustik - DAGA 2021,* Wien, Deutsche Gesellschaft für Akustik e. V., 2021, pp. 406-408.

[106] C. Rasmussen and C. Williams, Gaussian Processes for Machine Learning, Cambridge: The MIT Press, 2006.

[107] "Modeling auditory processing of amplitude modulation. I. detection and masking with narrow-band carriers," *Journal of the Acoustical Society of America 102 ,* pp. 2892-2905, 1997.

[108] R. Kruse, C. Borgelt, C. Braune, F. Klawonn, C. Moewes and M. Steinbrecher, "Allgemeine neuronale Netze," in *Computational Intelligence,* Wiesbaden, Springer Vieweg, 2011, pp. 33-42.

[109] S. Hochreiter and J. Schmidhuber, "Long Short-Term Memory," *Neural Computation 9 (8),* pp. 1735-1780, 1997.

[110] S. Siami-Namini, N. Tavakoli and A. Siami Namin, "The Performance of LSTM and BiLSTM in Forecasting Time Series," in *2019 IEEE International Conference on Big Data (Big Data),* Los Angeles, Institute of Electrical and Electronics Engineers, 2019, pp. 3285-3292.

[111] N. Mughees, S. A. Mohsin, A. Mughees and A. Mughees, "Deep sequence to sequence Bi-LSTM neural networks for day-ahead peak load forecasting," *Expert Systems with Applications 175 ,* p. 114844, 2021.

[112] Z. Xiang, J. Yan and I. Demir, "A Rainfall-Runoff Model With LSTM-Based Sequence-to-Sequence Learning," *Water resources research 56 (1),* 2020.

[113] D. Hadzalic, "Application of neural networks for prediction of subjectively assessed interior aircraft noise," KTH, School of Engineering Sciences (SCI), Aeronautical and Vehicle Engineering, Marcus Wallenberg Laboratory MWL, Stockholm, 2018.

[114] R. Imaizumi, R. Masumura, S. Shiota and H. Kiya, "Sequence-To-One Neural Networks for Japanese Dialect Speech Classification," in *2020 IEEE 9th Global Conference on Consumer Electronics (GCCE),* Kobe, Institute of Electrical and Electronics Engineers, 2020, pp. 933-935.

[115] M. A. Wani, F. A. Bhat, S. Afzal and A. I. Khan, Advances in Deep Learning, Warsaw: Springer, 2020.

[116] N. Srivastava, G. Hinton, A. Krizhevsky, I. Sutskever and R. Salakhutdinov, "Dropout: a simple way to prevent neural networks from overfitting," *Journal of Machine Learning Research 15*, pp. 1929-1958, 2014.

[117] J. Snoek, H. Larochelle and R. P. Adams, "Practical Bayesian Optimization of Machine Learning Algorithms," *Advances in neural information processing systems 25*, 2012.

[118] Y. Dokuz and Z. Tüfekçi, "Investigation of the Effect of LSTM Hyperparameters on Speech Recognition Performance," *European Journal of Science and Technology*, pp. 161-168, 2020.

[119] A. Weigend, "On overfitting and the effective number of hidden units," in *Proceedings of the 1993 connectionist models summer school*, Hillsdale, Lawrence Erlbaum Associates, 1994, pp. 335-342.

[120] S. Siami-Namini and A. Siami Namin, "Forecasting economics and financial time series: ARIMA vs. LSTM," *arXiv preprint arXiv: 1803.06386*, 2018.

[121] M. Biehl and H. Schwarze, "Learning by on-line gradient descent," *Journal of Physics A: Mathematical and general 28 (3)*, pp. 643-656, 1995.

[122] S. L. Özesmi, C. O. Tan and U. Özesmi, "Methodological issues in building, training, and testing artificial neural networks in ecological applications," *Ecological Modelling 195 (1-2)*, pp. 83-93, 2006.

A   Complete Results for the Electric Motor Order Variation Experiment

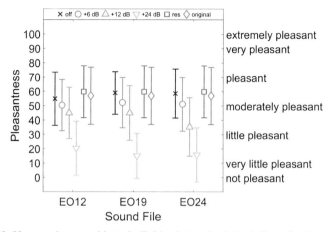

Figure 43. Mean values and interindividual standard deviations for the pleasantness of the electric motor order variation experiment for the original sound 1 recorded in a run-up condition with an initial speed of 0 km/h. The abscissa shows the order, whose level was modified. The colored symbols show the magnitude of the level variation, where "off" refers to a removal of the order, "res" refers to the residual sound, when all orders were removed and "original" refers to the original sound without variations.

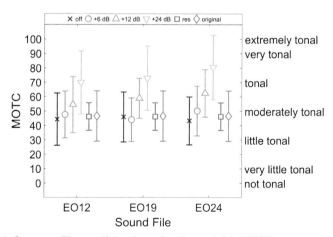

Figure 44. Same as Figure 43 but here for the variable MOTC.

135

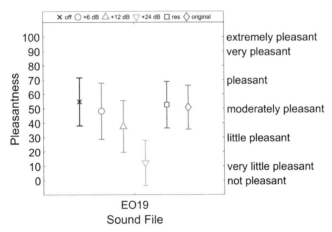

Figure 45. Same as Figure 43 but here for original sound 2 recorded in a run-up condition with an initial speed of 0 km/h.

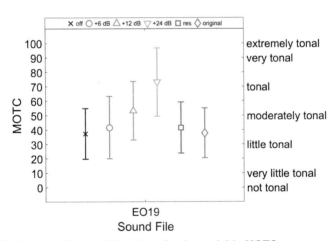

Figure 46. Same as Figure 45 but here for the variable MOTC.

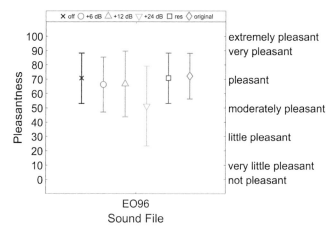

Figure 47. Same as Figure 43 but here for original sound 3 recorded in a run-up condition with an initial speed of 0 km/h. Note that for this vehicle only the 96th electric motor order was extractable. Therefore, the data points for "off" and "res" are the same.

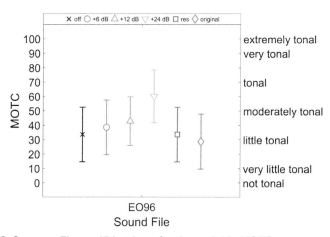

Figure 48. Same as Figure 47 but here for the variable MOTC.

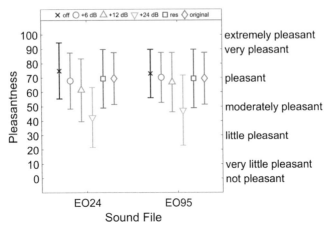

Figure 49. Same as Figure 43 but here for original sound 4 recorded in a run-up condition with an initial speed of 0 km/h.

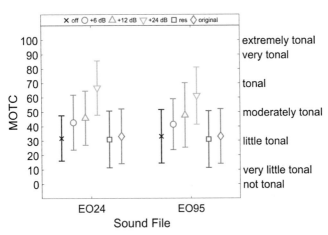

Figure 50. Same as Figure 49 but here for the variable MOTC.

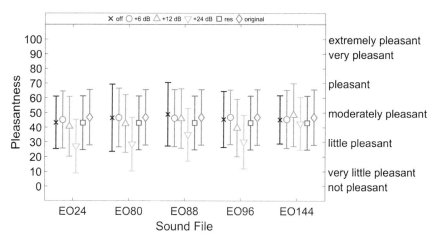

Figure 51. Same as Figure 43 but here for original sound 5 recorded in a run-up condition with an initial speed of 30 km/h.

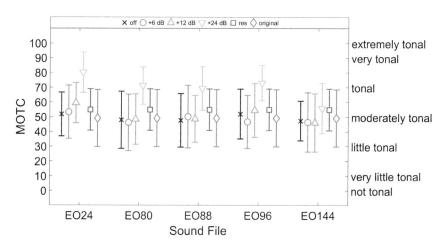

Figure 52. Same as Figure 51 but here for the variable MOTC.

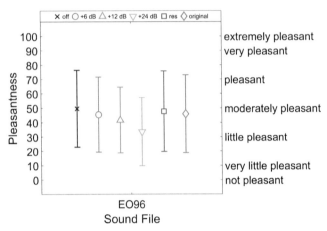

EO96
Sound File

Figure 53. Same as Figure 43 but here for original sound 6 recorded in a full-load run-up condition with an initial speed of 0 km/h. Note that for this vehicle only the 96th electric motor order is extractable. Therefore, the data points for "off" and "res" are the same.

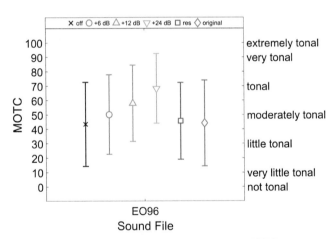

EO96
Sound File

Figure 54. Same as Figure 53 but here for the variable MOTC.

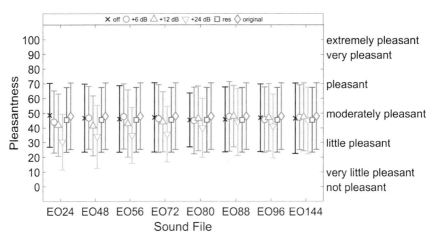

Figure 55. Same as Figure 43 but here for original sound 7 recorded in a full-load run-up condition with an initial speed of 50 km/h. This figure is also shown as Figure 31 in section 6.1.2.

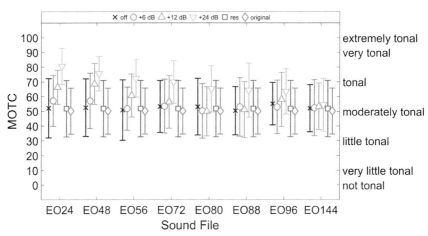

Figure 56. Same as Figure 55 but here for the variable MOTC. This figure is also shown as Figure 32 in section 6.1.2.

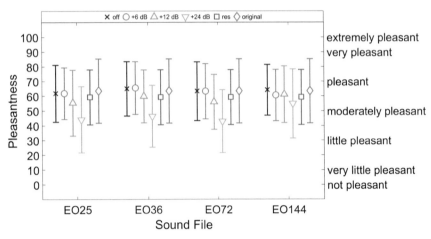

Figure 57. Same as Figure 43 but here for original sound 8 recorded in a coast-down condition with an initial speed of 50 km/h.

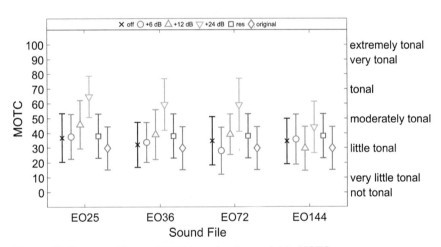

Figure 58. Same as Figure 57 but here for the variable MOTC.

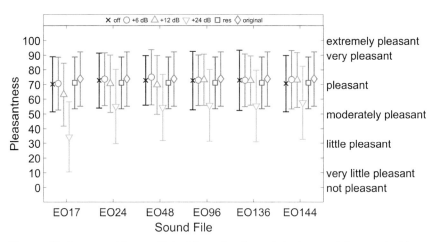

Figure 59. Same as Figure 43 but here for the original sound 9 recorded in a coast-down condition with an initial speed of 30 km/h.

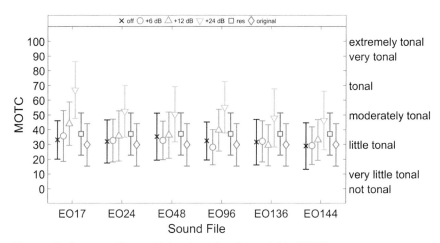

Figure 60. Same as Figure 59 but here for the variable MOTC.

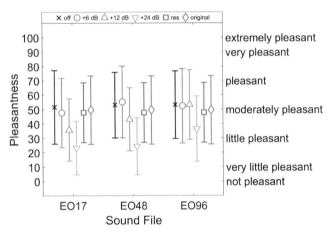

Figure 61. Same as Figure 43 but here for the original sound 10 recorded in a coast-down condition with an initial speed of 70 km/h.

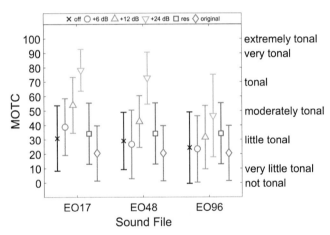

Figure 62. Same as Figure 61 but here for the variable MOTC.

B Complete Results for the Inverter Component Variation Experiment

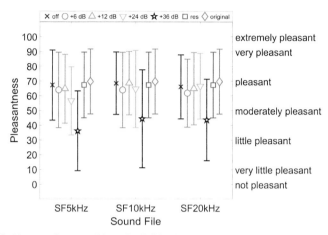

Figure 63. Mean values and interindividual standard deviations for the pleasantness of the inverter component level variation experiment for the original sound 1 recorded in a run-up condition with an initial speed of 0 km/h. The abscissa shows the inverter component, whose level was modified. The colored symbols show the magnitude of the level variation, where "off" refers to a removal of the component, "res" refers to the residual sound, when all components were removed and "original" refers to the original sound without variations.

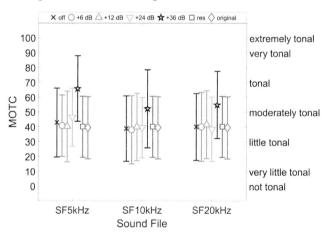

Figure 64. Same as Figure 63 but here for the variable MOTC.

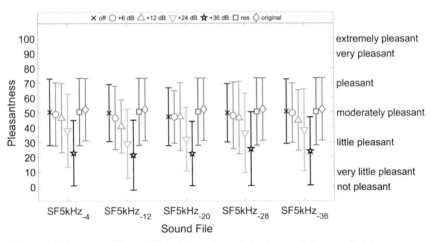

Figure 65. Same as Figure 63 but here for original sound 2 recorded in a run-up condition with an initial speed of 30 km/h. The variations of original sound 2 are split into two figures. The other variations are shown in Figure 67.

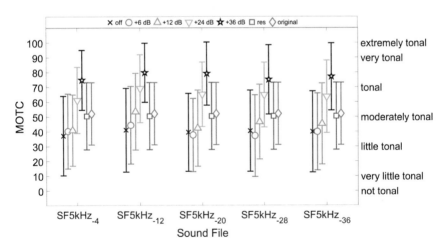

Figure 66. Same as Figure 65 but here for the variable MOTC. The variations of original sound 2 are split into two figures. The other variations are shown in Figure 68.

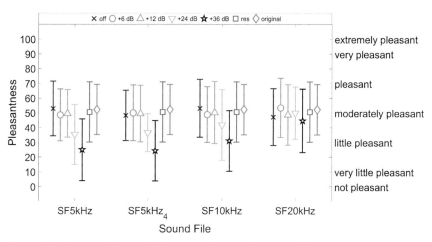

Figure 67. Same as Figure 63 but here for original sound 2 recorded in a run-up condition with an initial speed of 30 km/h. The variations of original sound 2 are split into two figures. The other variations are shown in Figure 65.

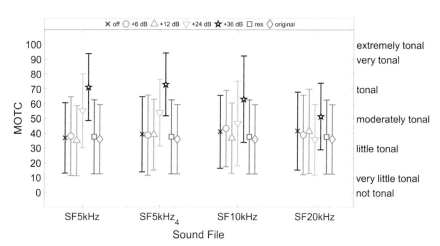

Figure 68. Same as Figure 67 but here for the variable MOTC. The variations of original sound 2 are split into two figures. The other variations are shown in Figure 66.

147

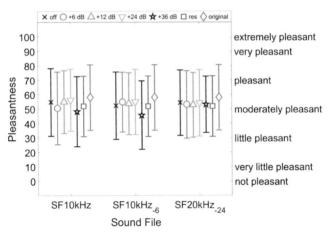

Figure 69. Same as Figure 63 but here for original sound 3 recorded in a full-load run-up condition with an initial speed of 0 km/h.

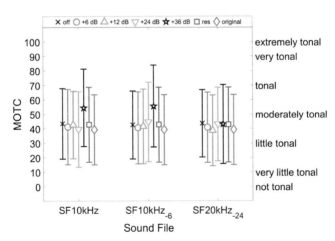

Figure 70. Same as Figure 69 but here for the variable MOTC.

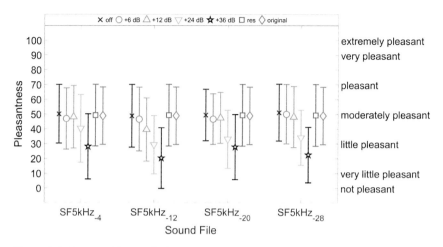

Figure 71. Same as Figure 63 but here for original sound 4 recorded in a full-load run-up condition with an initial speed of 30 km/h. The variations of original sound 4 are split into two figures. The other variations are shown in Figure 73.

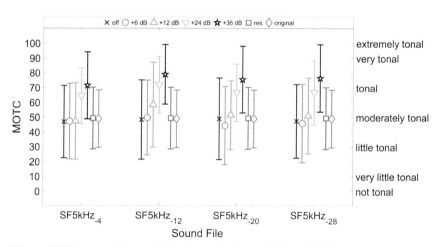

Figure 72. Same as Figure 71 but here for the variable MOTC. The variations of original sound 4 are split into two figures. The other variations are shown in Figure 74.

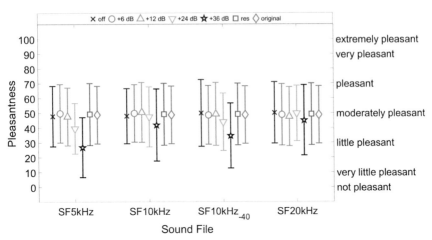

Figure 73. Same as Figure 63 but here for original sound 4 recorded in a full-load run-up condition with an initial speed of 30 km/h. The variations of original sound 4 are split into two figures. The other variations are shown in Figure 71.

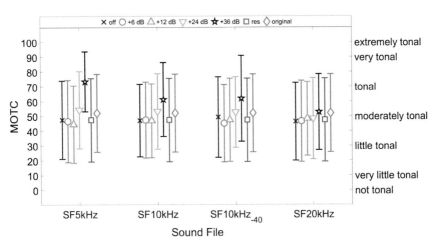

Figure 74. Same as Figure 73 but here for the variable MOTC. The variations of original sound 4 are split into two figures. The other variations are shown in Figure 72.

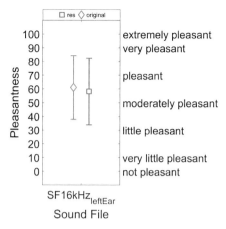

Figure 75. Same as Figure 63 but here for original sound 5 recorded in a coast-down condition with an initial speed of 50 km/h. Note that from this sound only from the left ear channel one order could be extracted and only the level of binaurally extractable inverter components was varied.

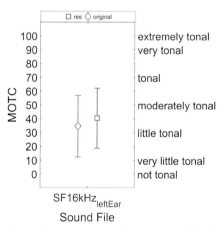

Figure 76. Same as Figure 75 but here for the variable MOTC.

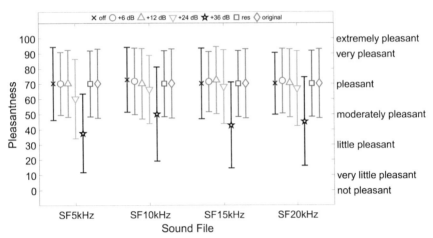

Figure 77. Same as Figure 63 but here for original sound 6 recorded in a coast-down condition with an initial speed of 30 km/h.

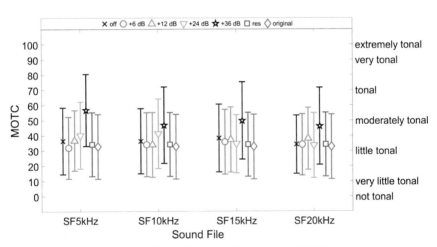

Figure 78. Same as Figure 77 but here for the variable MOTC.

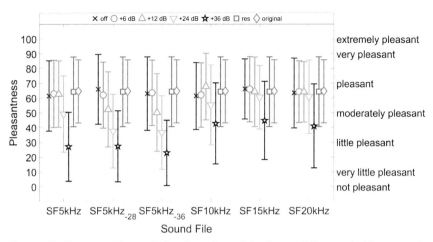

Figure 79. Same as Figure 63 but here for original sound 7 recorded in a coast-down condition with an initial speed of 50 km/h. The figure is also shown as Figure 33 in section 6.2.2.

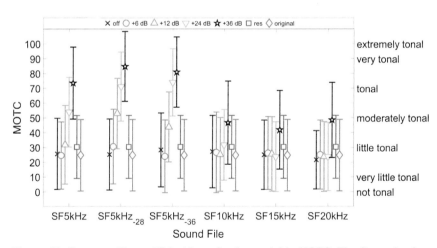

Figure 80. Same as Figure 79 but here for the variable MOTC. The figure is also shown as Figure 34 in section 6.2.2.

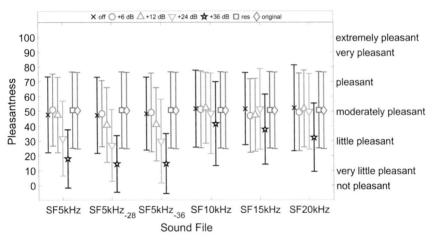

Figure 81. Same as Figure 63 but here for original sound 8 recorded in a coast-down condition with an initial speed of 70 km/h.

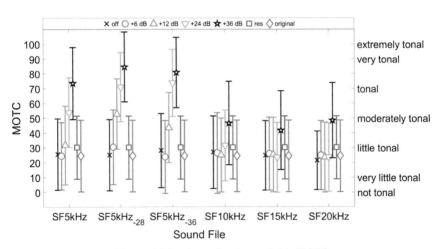

Figure 82. Same as Figure 81 but here for the variable MOTC.